Simulation of Complex Systems

Simulation of Complex Systems

Aykut Argun, Agnese Callegari and Giovanni Volpe
Department of Physics, University of Gothenburg, Göteborg, Sweden

IOP Publishing, Bristol, UK

ISBN 978-0-7503-3843-1 (ebook)
ISBN 978-0-7503-3841-7 (print)
ISBN 978-0-7503-3844-8 (myPrint)
ISBN 978-0-7503-3842-4 (mobi)

DOI 10.1088/978-0-7503-3843-1

Version: 20212001

IOP ebooks

British Library Cataloguing-in-Publication Data: A catalogue record for this book is available from the British Library.

Published by IOP Publishing, wholly owned by The Institute of Physics, London

IOP Publishing, Temple Circus, Temple Way, Bristol, BS1 6HG, UK

US Office: IOP Publishing, Inc., 190 North Independence Mall West, Suite 601, Philadelphia, PA 19106, USA

Contents

Preface x

Author biography xv

1 Molecular dynamics **1-1**

1.1 Single particle 1-2
1.2 Time reversibility 1-7
1.3 Multiple particles 1-8
1.4 Randomness 1-11
1.5 Further reading 1-13
1.6 Problems 1-14
1.7 Challenges 1-16
References 1-17

2 Ising model **2-1**

2.1 Monte Carlo method 2-2
2.2 Ising model 2-3
2.3 Critical temperature 2-6
2.4 Critical mixtures 2-8
2.5 Further reading 2-10
2.6 Problems 2-10
2.7 Challenges 2-11
References 2-12

3 Forest fires **3-1**

3.1 Forest growth and fire ignition 3-2
3.2 Power-law behavior 3-3
3.3 Further reading 3-9
3.4 Problems 3-9
3.5 Challenges 3-10
References 3-11

4 The Game of Life **4-1**

4.1 One-dimensional cellular automata 4-2
4.2 Conway's Game of Life 4-5

4.3	Majority rule	4-9
4.4	Further reading	4-12
4.5	Problems	4-12
4.6	Challenges	4-13
	References	4-14
5	**Brownian dynamics**	**5-1**
5.1	Random walks and universality	5-2
5.2	Discrete white noise	5-3
5.3	Brownian motion	5-6
5.4	Optical tweezers	5-7
5.5	Further reading	5-9
5.6	Problems	5-10
5.7	Challenges	5-11
	References	5-12
6	**Anomalous diffusion**	**6-1**
6.1	Anomalous diffusion exponent	6-2
6.2	Regularization and normalization	6-4
6.3	Models of anomalous diffusion	6-5
6.4	Anomalous diffusion in a non-homogeneous force field	6-8
6.5	Further reading	6-11
6.6	Problems	6-11
6.7	Challenges	6-12
	References	6-13
7	**Multiplicative noise**	**7-1**
7.1	A minimal discrete-time model	7-2
7.2	Position-dependent noise	7-4
7.3	Stochastic integrals	7-6
7.4	The spurious drift	7-7
7.5	Drift and diffusion measurement	7-8
7.6	Particles close to an interface	7-10
7.7	Further reading	7-13
7.8	Problems	7-13
7.9	Challenges	7-14
	References	7-14

8	**The Vicsek model**	**8-1**
8.1	The standard Vicsek model	8-2
8.2	The effect of delay	8-8
8.3	Non-metric and non-reciprocal interactions	8-10
8.4	Further reading	8-11
8.5	Problems	8-12
8.6	Challenges	8-13
	References	8-13
9	**Living crystals**	**9-1**
9.1	Active Brownian motion	9-2
9.2	Mean square displacement	9-4
9.3	Living crystals	9-5
9.4	Aligning interactions	9-8
9.5	Further reading	9-11
9.6	Problems	9-12
9.7	Challenges	9-13
	References	9-13
10	**Sensory delay**	**10-1**
10.1	A light-sensitive robot	10-2
10.2	Single robot with a sensory delay	10-4
10.3	Multiple robots with sensory delay	10-6
10.4	Further reading	10-8
10.5	Problems	10-8
10.6	Challenges	10-9
	References	10-9
11	**Disease spreading**	**11-1**
11.1	The agent-based SIR model	11-2
11.2	Disease transmission as a function of the infection rate	11-4
11.3	Extended SIR models	11-5
11.4	Lockdown strategies	11-7
11.5	Further reading	11-8
11.6	Problems	11-8
11.7	Challenges	11-9
	References	11-9

12	**Network models**	**12-1**
12.1	The adjacency matrix	12-2
12.2	Path length, diameter, and clustering coefficient	12-5
12.3	Erdős–Rényi random graphs	12-6
12.4	Watts–Strogatz small-world graphs	12-7
12.5	Albert–Barabási preferential-growth graphs	12-9
12.6	Further reading	12-10
12.7	Problems	12-10
12.8	Challenges	12-11
	References	12-11
13	**Evolutionary games**	**13-1**
13.1	The prisoner's dilemma	13-2
13.2	Evolutionary games on a lattice	13-3
13.3	Multiple strategies	13-7
13.4	Further readings	13-8
13.5	Problems	13-8
13.6	Challenges	13-9
	References	13-10
14	**Ecosystems**	**14-1**
14.1	Lotka–Volterra model	14-2
14.2	The logistic growth model	14-10
14.3	Mutualism	14-11
14.4	Competition	14-13
14.5	Further reading	14-14
14.6	Problems	14-14
14.7	Challenges	14-16
	References	14-16
15	**Ant-colony optimization**	**15-1**
15.1	The minimum path length problem	15-2
15.2	Ants at work	15-6
15.3	Interruptions, accidents, and randomness	15-10
15.4	Further reading	15-12
15.5	Problems	15-13

15.6 Challenges 15-14
References 15-15

16 The Sugarscape 16-1

16.1 Models of segregation 16-2
16.2 The Sugarscape 16-5
16.3 Further reading 16-11
16.4 Problems 16-12
16.5 Challenges 16-13
References 16-14

Preface

Complex systems are found everywhere. The burst of neuronal activity when you wake up in the morning is a result of complex activation waves in the intricate neuronal network constituting your brain. The coffee you drink at breakfast is delivered to you by a complex logistics network extending all over the Earth. When you leave your house to go to work, you need to navigate a complex public transport system. At work you might meet with different colleagues, who are part of a complex organization, such as a university or a company. When you slack off on your favorite social network app, you receive a flow of information that depends on your place in the social network—for example, your closest contacts and your communities. If you are unlucky in your face-to-face interactions, you might get a communicable disease, and the probability of this happening depends on the physical and social characteristics of your environment.

Clearly, there are plenty of complex systems all around us. But, what is a 'complex system'? How can it be defined?

A complex system can be defined as a system composed of many interacting parts that features emergent behavior. Here, the keywords are 'many', 'interacting', and 'emergent'.

The first property of a complex system is that it is constituted of *many parts*. For example, we can think of the many individuals in a social network, the many species in an ecosystem, or the many neurons in a brain.

The second property is that the parts in a complex system must *interact* with each other. For example, the individuals in a social network might be sending messages or sharing news with each other. In an ecosystem, some species hunt, others are hunted, others again compete for the same resources or establish mutualistic relationships. The neurons in a brain are connected by synapses, so that one neuron can excite or inhibit other neurons.

The third property is that the interaction between the many parts of a complex system must lead to some non-trivial *emergent* behaviors. For example, in social networks, we can see the emergence of a steady flow of (mis)information, which leads to complex social and political dynamics. The introduction of a new species into an ecosystem can lead to coexistence or extinction, depending on many subtle conditions. The excitation and inhibition between neurons in a brain leads to complex patterns of neuronal activity, which are the foundation of our consciousness.

In order to be able to gain insights into complex systems, we need appropriate tools. One of the most powerful tools is provided by numerical simulations, which are nowadays within reach of anyone with a laptop—hence the title and topic of this book: 'Simulation of Complex Systems'.

Historically, the tools employed to describe the world have determined the focus of scientific inquiry, while also delimiting the boundaries of what is knowable or worth knowing. Over the course of Western civilization, we have gone through three phases. Each shift has offered unprecedented opportunities to study new phenomena and gain new insights, but has also posed unexpected practical and epistemological challenges.

The first phase was rooted in *geometry*. In fact, geometry remained at the core of science from the time of the ancient Greeks until the Renaissance. One of the reasons for this was the success of Euclid's 'Elements', which is still seen as one of the greatest intellectual achievements of humanity, and of Ptolemaic celestial mechanics, which accurately predicted the motion of celestial bodies, an essential for maritime navigation. Because of the primacy of geometry, a good research question was supposed to be framed in geometrical terms. An example of a good question was how to draw a regular heptagon using a compass and a straightedge. On the other hand, irrational and imaginary numbers were inconceivable, because of the lack of an acceptable geometrical interpretation.

We have to wait until the 19th century to move past geometry into the second phase, when, thanks to the continued success of Kepler's laws and Newtonian mechanics, *calculus* became prominent. The focus shifted toward describing physical phenomena using equations—ideally simple, mathematically elegant equations. An example of a good research question was now how to describe the trajectory of a projectile (or of the legendary apple inspiring Newton's theory). A less welcome question was that of how to explore the conditions under which a deterministic system becomes unpredictable, for example, in the case of the weather or the three-body problem.

It was only from the middle of the last century that we moved into the third phase. Thanks to the invention of computers, *numerical simulations* have now become more and more commonplace, first in engineering and later also in the basic sciences; simulations of airfoils around airplanes and of black holes are widely regarded as solid science, and very recently, even mathematicians have started to accept computational proofs of mathematical theorems.

Of course, this is not the end of the story (it never is), as numerical simulations still require an algorithm written by a human. We can already see a fourth phase approaching on the horizon, a phase in which algorithms will be written by machines—but that is a topic for a follow-up book!

This book embraces the use of numerical simulations to understand complex systems. Thus, we will explain the effective numerical simulation techniques most often used to approach a variety of complex systems that are of fundamental importance in physics, biology, engineering, social sciences, and economics.

Beyond the intrinsic interest aroused by the use of numerical simulations for the modelling and understanding of phenomena for applications, numerical simulations are ideal tools with which to gain direct, hands-on experience of complex systems. Furthermore, numerical experiments are inexpensive and within the reach of anyone with access to a computer.

We have written this book at a master's degree and graduate student level, making it ideal for a course on the simulation of complex system as well as for self-study. We would like to remark at this point that this is not a programming or physics book, so we assume a solid knowledge of programming (regardless of the programming language) as well as basic physics (mechanics and statistical physics at the undergraduate level should suffice).

Each chapter is an independent, largely self-contained class on a different topic. Each chapter includes an introduction and the motivation for the topic, a description of relevant numerical approaches to the problem at hand with guided *exercises* (these are an integral part of the learning process and should be solved by every reader who wants to acquire a deep understanding of the technique), a list of further reading, some *problems* extending the reach of the exercises (which should be feasible using the techniques studied in the chapter with a bit of dedicated effort), and some *challenges* involving more advanced research questions (these are more difficult than the problems, sometimes reaching the level of open research questions). With this training, it should be possible for readers to reach a level sufficient to perform some original research in these fields and even to publish some original articles of their own. Of course, this will require a lot of dedicated effort and a thorough study of the relevant scientific literature beyond the further reading proposed in this book—but we hope to take readers to the point where they can confidently embark on this journey.

Since the chapters are largely independent, they can be read in any order and, of course, the reader can also focus on just a subset of the chapters. We have nevertheless tried to arrange the topics in a logical order.

We start with some basic topics, which describe numerical simulation techniques that are of fundamental importance in physics and engineering (e.g. molecular dynamics, passive and active Brownian dynamics, anomalous diffusion, and multiplicative noise) and continue with topics that are of more specialized interest in biology, engineering, and the social sciences (e.g., the Game of Life, the Vicsek model, the spread of disease, network models, communities, ecosystems, evolutionary games, ant-colony optimization, the Sugarscape):

Chapter 1. Molecular dynamics. Molecular dynamics are used to explore the thermodynamics of systems of interest in physics and biology. Furthermore, learning how to simulate molecular dynamics provides a strong foundation with which to understand simulations based on differential equations in general.

Chapter 2. Ising model. The Ising model provides a very versatile framework for a wide range of phenomena related to phase transitions, such as ferromagnetism and magnetization. Furthermore, studying the properties of the Ising model provides insights into critical phenomena, critical fluctuations, and critical Casimir forces.

Chapter 3. Forest fires. Modeling the growth of a forest that experiences the repeated occurrence of fires provides a natural tool with which to introduce concepts related to power-law distributions and self-organized criticality, which can be used to understand a broad range of phenomena.

Chapter 4. The Game of Life. Cellular automata are very powerful model systems used to study the emergence of complexity and its statistical properties. In particular, Conway's Game of Life is the most famous example of a cellular automaton, in which a set of simple deterministic rules operating in a simplified environment can lead to very complex behaviors—reminiscent of life.

Chapter 5. Brownian dynamics. Brownian dynamics is a fundamental numerical technique used to simulate the Brownian motion of microscopic particles and to explore the physics of microscopic systems. It also provides key insights into how to numerically solve stochastic differential equations, which have a broad range of applications—from ecology to finance.

Chapter 6. Anomalous diffusion. Anomalous diffusion goes beyond the standard Brownian diffusion of microscopic particles, and typically happens in systems that are driven out of thermodynamic equilibrium or those that are in complex environments. It is important in describing several phenomena of physical and biological interest, such as the intracellular transport of chemicals and the foraging behaviors of animals.

Chapter 7. Multiplicative noise. Multiplicative noise occurs when the noise driving a system depends on the state of the system, for example, when the diffusion of a particle depends on the particle's position. It plays an important and subtle role in a wide range of phenomena, from signal processing to population dynamics. Modeling multiplicative noise correctly is tricky and has subtle connections to the definition of stochastic integrals.

Chapter 8. The Vicsek model. The emergence of collective phenomena, such as swarming, occurs in many biological systems, such as flocks of birds, schools of fish, and swarming bacteria. The Vicsek model is a classical model used to study the collective motion that emerges from local interactions between motile individuals. It has also been extended to account for non-metric and non-reciprocal interactions.

Chapter 9. Living crystals. Active Brownian motion is a very powerful model that describes the active motion of microscopic self-propelling particles, such as motile bacteria and artificial micromotors. When multiple active particles interact, they can give rise to interesting complex phenomena, such as metastable clusters (aka 'living crystals') and metastable channels.

Chapter 10. Sensory delay. In many real systems, sensory delays occur between the time at which a stimulus is received and when the ensuing action is performed. Such delays can greatly alter the way in which an agent (or a group of agents) behaves. Furthermore, sensory delays can be used as design parameters to control the agents' behavior.

Chapter 11. Disease spreading. The spread of infectious disease in a population can have major ecological, societal, and economic consequences. For this reason, several models have been developed to predict the spread of disease and to design optimal policies to control it.

Chapter 12. Network models. Networks (or graphs) are central in many fields, from physics and the neurosciences to ecology and sociology. Networks can be characterized by different measures. Several different network models have been proposed to explain the properties of networks found in real life.

Chapter 13. Evolutionary games. Game theory is the study of how to get the best outcome in situations in which multiple agents are interacting. Several paradoxes are driven by the fact that the best outcome for the individual and that for the group are often at odds—the most famous being the prisoner's dilemma. Things become even more interesting when agents are allowed to play multiple times and evolve more beneficial strategies.

Chapter 14. Ecosystems. Complex interspecies relations emerge in ecosystems, including predation, mutualism, and competition. These can be modeled using differential equations, such as the logistic equation and the famous Lotka–Volterra model.

Chapter 15. Ant-colony optimization. Ant-colony optimization is a technique for solving computational problems related to finding paths through graphs, which is inspired by the strategies that real ants use to share vital information (e.g. the locations of food sources) with fellow ants. This is a heuristic method that has proven extremely effective for numerous complex problems in multiple fields.

Chapter 16. The Sugarscape. The Sugarscape is an agent-based social simulation, in which individuals (agents) and the environment (the sugar-producing landscape) interact through a set of rules. Similar models are often employed in sociology, for example, to understand the evolution of the social fabric of a society.

Each chapter is also complemented by example Python scripts and a discussion forum, which can be found at https://github.com/softmatterlab/SOCS/ (the GitHub page associated with this book).

Author biography

Aykut Argun

Aykut Argun is a researcher at the University of Gothenburg, Sweden. His PhD in Physics was awarded by the University of Gothenburg in 2021. His research interests are optical trapping and manipulation, statistical physics, soft matter, active matter, and machine learning as applied to experimental data analysis. He developed the software packages DeepTrack—deep learning for digital microscopy, DeepCalib—deep learning to calibrate force fields and RANDI—deep learning for the characterization of anomalous diffusion.

Agnese Callegari

Agnese Callegari is a researcher at the University of Gothenburg, Sweden. Her PhD was awarded by the University of Rome Tor Vergata in 2003. Her research interests are optical trapping and manipulation, statistical physics, soft matter, and active matter. She has authored more than ten publications on these topics. She contributed to the development of the software package OTGO—optical tweezers in geometrical optics.

Giovanni Volpe

Giovanni Volpe is Professor at the Physics Department of the University of Gothenburg, Sweden, where he leads the Soft Matter Lab. His research interests include soft matter, optical trapping and manipulation, statistical mechanics, brain connectivity, and machine learning. He has authored more than 100 articles and reviews on soft matter, statistical physics, optics, physics of complex systems, brain network analysis, and machine learning. He co-authored the book 'Optical Tweezers: Principles and Applications' (Cambridge University Press, 2015). He has developed several software packages (OTS—the optical tweezers software, Braph—brain analysis using graph theory, DeepTrack—deep learning for digital microscopy, and DeepCalib—deep learning to calibrate force fields).

IOP Publishing

Simulation of Complex Systems

Aykut Argun, Agnese Callegari and Giovanni Volpe

Chapter 1

Molecular dynamics

Matter is made of *microscopic* components, such as atoms and molecules. In a very simplified picture, we can imagine that atoms are like small spheres that have a mass and possibly an electric charge, which interact with each other to form the *macroscopic* objects we see in Nature. Even though complex and advanced concepts based on quantum mechanics are required to derive the properties of matter from *first principles*, it is often convenient and sufficient to employ simpler *effective models* based on classical mechanics and Newton's laws of motion.

Figure 1.1. Solar system. Molecular dynamics can be used to simulate a broad range of microscopic and macroscopic systems. For example, they can simulate the motion of the planets in the Solar System under the influence of gravitational forces. *Source: NASA. Artistic representation of the Solar System showing the Sun, inner planets, asteroid belt, outer planets, and a comet.* Courtesy of NASA.

Let us consider a simple example, in which we can derive macroscopic physical concepts using only Newton's laws: the *ideal gas*. An ideal gas is a collection of identical, non-interacting point-like particles with a fixed mass, each traveling with its own velocity within an enclosed volume, typically a cubic container. While moving along its straight trajectory, each particle eventually hits a wall of the container and bounces back. This causes an exchange of momentum between the particle and the wall. By averaging the effect of many particle–wall collisions occurring over a certain time, we can define a *pressure*, which is a macroscopic quantity, and relate it to the microscopic properties of the motion of the particles.

The approach we have just illustrated for the ideal gas is applicable to many other situations and systems. Importantly, it can be straightforwardly applied to the simulation of complex systems—a technique known as *molecular dynamics*. For example, molecular dynamics has been used to simulate the dynamics of gases, liquids, and crystals. In chemistry and biology, it has been employed to refine the 3D structures, obtained by spectroscopic data, of polymers, proteins, and other biomolecules. It can even simulate macroscopic systems, such as the motion of the planets in the Solar System (figure 1.1). In each case, the particles (e.g., atoms, molecules, or planets) interact with each other through specific forces (e.g., electrostatic, van der Waals, or gravitational forces) following Newton's laws of classical mechanics.

In this chapter, we introduce the use of molecular dynamics to simulate the evolution of complex systems over time. We start from Newton's equations of motion for a single point-like particle and introduce the standard numerical algorithms used to solve them, discussing their advantages and limitations. We show how to preserve the time reversibility and energy conservation that we expect in real physical systems. We then extend these algorithms to more complex systems of many mutually-interacting point-like particles and discuss the most common interactions used in molecular dynamics. Finally, we present how apparent randomness emerges in a system of many deterministically-interacting particles, leading to *Brownian motion*, which will be discussed in more detail in chapter 5.

Example codes: Example Python scripts related to this chapter can be found at: https://github.com/softmatterlab/SOCS/tree/main/Chapter_01_Molecular_Dynamics. Readers are welcome to participate in the discussions related to this chapter at: https://github.com/softmatterlab/SOCS/discussions/7.

1.1 Single particle

Given a point-like particle of mass m, Newton's second law in one dimension states that

$$F = ma, \tag{1.1}$$

where F is the net force acting on the particle, and a is the particle acceleration. Given the initial position r_0 of the particle and its initial velocity v_0, equation (1.1) can be rewritten as follows:

$$\begin{cases} \dfrac{\mathrm{d}^2 r(t)}{\mathrm{d}t^2} & = \dfrac{F}{m} \\ r(0) & = r_0 \\ \dfrac{\mathrm{d}r}{\mathrm{d}t}(0) & = v_0 \end{cases} \tag{1.2}$$

where $r(t)$ is the particle position as a function of time t. This second-order differential equation can be usefully rewritten as a set of two first-order differential equations, introducing the instantaneous velocity $v(t)$ as an auxiliary variable:

$$\begin{cases} \dfrac{\mathrm{d}r(t)}{\mathrm{d}t} & = v(t) \\ \dfrac{\mathrm{d}v(t)}{\mathrm{d}t} & = \dfrac{F}{m} \\ r(0) & = r_0 \\ v(0) & = v_0 \end{cases} . \tag{1.3}$$

Some (often implicit) assumptions in classical mechanics are that time and space are continuous variables, and that they can be represented in a Cartesian space. We are therefore drawn to imagine the physical trajectory of a particle as a *continuous* line. In experiments and simulations, however, we have to consider *discrete* trajectories. For example, if we track the position of an object using a camera, we will only have access to its positions at the discrete times corresponding to the frames captured by the camera. Similarly, if we simulate a particle trajectory on a computer, we will only be able to generate a discrete sequence of positions:

$$\begin{array}{ccc} t_0 & \leftrightarrow & x_0 \\ t_1 & \leftrightarrow & x_1 \\ t_2 & \leftrightarrow & x_2 \\ \vdots & & \vdots \\ t_n & \leftrightarrow & x_n \\ t_{n+1} & \leftrightarrow & x_{n+1} \\ \vdots & & \vdots \end{array}$$

Usually, the time sequence t_0, t_1, ... is made from multiples of a time step Δt, i.e., $t_n = n \cdot \Delta t$. Typically, the smaller Δt, the more accurate the representation of the trajectory.

To simulate a trajectory, we proceed to sequentially determine the positions of the particle on a time step by time step basis—a procedure known as *time propagation*. The simplest method of time propagation that solves equation (1.3) is Euler's propagation scheme or, more simply, the *Euler algorithm*. We start from the Taylor series of the position and velocity truncated to the first order:

$$\begin{cases} r(t_0+\Delta t) &= r(t) + \left.\dfrac{dr(t)}{dt}\right|_{t_0} \Delta t + o(\Delta t) &= r(t_0) + v(t_0)\,\Delta t + o(\Delta t) \\ v(t_0+\Delta t) &= v(t) + \left.\dfrac{dv(t)}{dt}\right|_{t_0} \Delta t + o(\Delta t) &= v(t_0) + \dfrac{F(t_0)}{m}\,\Delta t + o(\Delta t) \end{cases} . \quad (1.4)$$

Ignoring the term $o(\Delta t)$, which represents the infinitesimals of order larger than one, we obtain the finite-difference integration scheme of the Euler algorithm:

$$\begin{cases} x_{n+1} &= x_n + v_n \Delta t \\ v_{n+1} &= v_n + \dfrac{F_n}{m}\Delta t \end{cases} \quad (1.5)$$

where the force $F_n = F(t_n, r_n, v_n)$ is calculated using the position and the velocity at time t_n. Generally, the force may depend on the particle position (e.g., elastic and gravitational forces), the particle velocity (e.g., viscous and dynamic friction forces), and also time (e.g., periodic forces). One of the simplest cases is a conservative force:

$$F(r) = -\nabla U(r), \quad (1.6)$$

which is a position-dependent force given by the gradient of a potential $U(r)$. In some cases, there is an analytical expression for $F(r)$, which makes the calculation of F_n straightforward. In other cases, however, it might be necessary to calculate F_n numerically, as follows:

$$F_n = F(r_n) = -\frac{U(r_n + \Delta r) - U(r_n - \Delta r)}{2\Delta r}, \quad (1.7)$$

where Δr is a positional increment small enough to ensure that the potential is locally linear as a function of position.

Exercise 1.1. Harmonic oscillator simulated using the Euler algorithm. Let us consider a harmonic oscillator: a point-like particle of mass m attached to a perfect spring with an elastic constant k. The particle is placed on a horizontal, frictionless surface. Initially, the particle is at position r_0 with velocity v_0.

a. Analytically calculate the trajectory of the particle as a function of time. Specifically, show that $r(t) = A\cos(\omega t + \phi)$, with $\omega = \sqrt{k/m}$, $A = \sqrt{r_0^2 + (v_0/\omega)^2}$, and ϕ defined by $\cos\phi = \frac{r_0}{A}$ and $\sin\phi = -\frac{v_0}{A\omega}$. Also, show that $v(t) = -\omega A \sin(\omega t + \phi)$.

b. Write a program that propagates the trajectory using the Euler algorithm. What is an appropriate value for Δt? *[Hint: Compare it with the characteristic frequency of the motion,* $f = \omega/(2\pi)$*.]*

c. Calculate the total mechanical energy of the particle as a function of time, analytically and by simulation. Compare the results. *[Hint: The total energy E is the sum of the potential energy U and the kinetic energy K, which, for a harmonic oscillator, are* $U(r) = \frac{1}{2}kr^2$ *and* $K(v) = \frac{1}{2}mv^2$*.]*

d. Is the total mechanical energy conserved by the analytical solution? And by the simulation?

While doing exercise 1.1, we note that the trajectory obtained using the Euler algorithm eventually deviates from the analytical solution, no matter how small we make Δt, as shown in figure 1.2. This is due to the fact that the force is calculated using the particle position at the beginning of each time interval $I_n = [t_n, t_{n+1}]$, leading to a systematic error in the estimate of the velocity change. This systematic error is particularly evident in the case of the harmonic oscillator (can you figure out why?).

A much better approximation is obtained by using the particle position at the middle of the time interval I_n to calculate the force. This is what the *leapfrog algorithm* does: the position is first advanced for half a time step to obtain $r_{n+\frac{1}{2}}$, then the velocity v_{n+1} is calculated using $F_{n+\frac{1}{2}} = F(r_{n+\frac{1}{2}})$, and finally the position is advanced for another half time step to r_{n+1}:

$$\begin{cases} r_{n+\frac{1}{2}} &= r_n + v_n \dfrac{\Delta t}{2} \\ v_{n+1} &= v_n + \dfrac{F_{n+\frac{1}{2}}}{m} \Delta t \\ r_{n+1} &= r_{n+\frac{1}{2}} + v_{n+1} \dfrac{\Delta t}{2} \end{cases} . \tag{1.8}$$

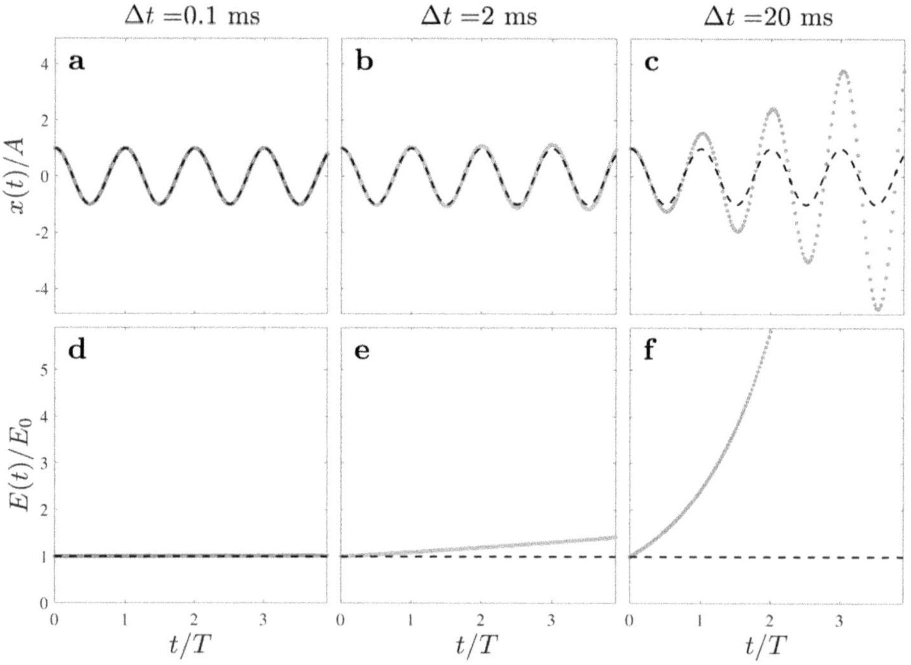

Figure 1.2. Euler algorithm. (a)–(c) Trajectories and (d)–(f) total mechanical energy for a harmonic oscillator simulated using the Euler algorithm for different time steps Δt. In all cases, the numerical trajectory ((a) blue, (b) green, and (c) orange dots) eventually deviates from the analytical solution (black dashed line in (a)–(c)), while the total mechanical energy of the simulation (blue (d), green (e), and orange (f) dots) is not constant as predicted by the analytical solution (black dashed line in (d)–(f)). These deviations increase for longer Δt. Note that even for small time steps ((a) and (d)) the error eventually causes divergence. The simulation parameters are $m = 0.1$ kg, $k = 5$ N m^{-1}, $x_0 = 0.1$ m, and $v_0 = 0$ m s^{-1}.

Exercise 1.2. Leapfrog algorithm propagator. Show that the propagator of the leapfrog algorithm given in equation (1.8) can be written as a function of only x_n, v_n, F_n, m, and Δt as

$$x_{n+1} = x_n + v_n \cdot \Delta t + \frac{1}{2} \cdot \frac{F_n}{m} \cdot \Delta t^2. \tag{1.9}$$

Show also that this algorithm is $o(\Delta t^2)$. *[Hint: Use the Taylor expansion.]*

Exercise 1.3. Harmonic oscillator simulated using the leapfrog algorithm. Repeat exercise 1.1 using the leapfrog algorithm instead of the Euler algorithm.

a. Write a program that simulates the trajectory using the leapfrog algorithm.

b. Compare the simulated trajectory with the analytical trajectory and with the trajectory simulated using the Euler method. What do you observe?

c. Compare the total energies of the analytical and numerical solutions. Which method appears to perform better?

While completing exercise 1.3, it becomes evident that the leapfrog method does not introduce drift in the mechanical energy as the Euler algorithm did (exercise 1.1). The corresponding results are shown in figure 1.3.

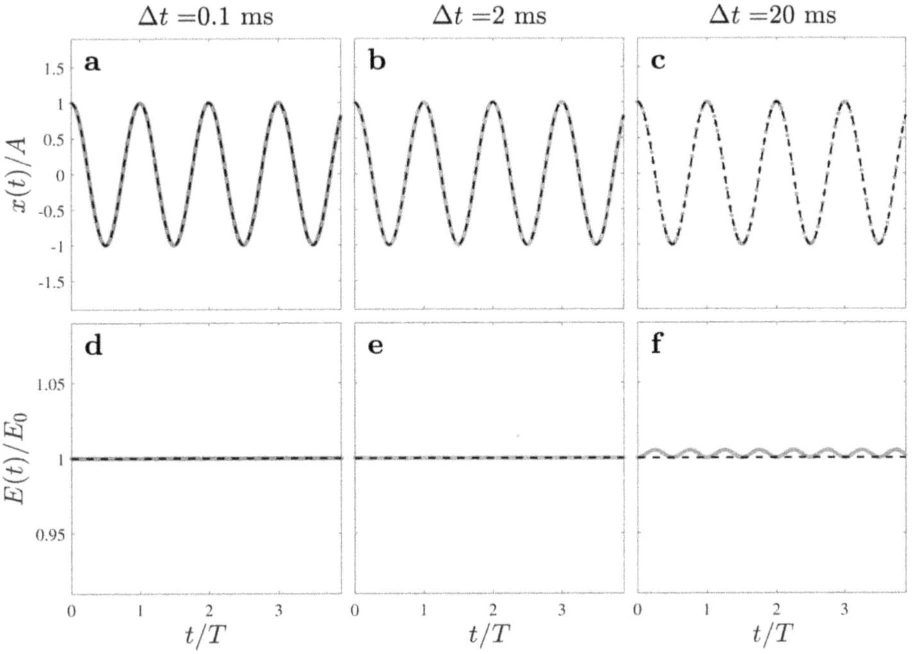

Figure 1.3. Leapfrog algorithm. Simulation of the same system as that shown in figure 1.2 using the leapfrog algorithm instead of the Euler algorithm. Note how, in this case, there is good agreement between the simulations and the analytic solution, and that mechanical energy is conserved (albeit with slight fluctuations) for all time steps Δt.

1.2 Time reversibility

Energy conservation is related to the concept of *time reversibility*. Newton's equations, and in fact the whole of classical physics, are time reversible. This means that if the sign of the velocity of a particle is changed at a certain time, the solution exactly backtracks the trajectory in phase space. It is therefore important that molecular dynamics algorithms produce trajectories that are time reversible. In a numerical simulation, time reversibility can be checked in the following way: first, generate a trajectory by obtaining a particle's position and velocity for a number of time steps N; then, invert the sign of all velocities and repeat the simulation for another N time steps; if the new trajectory backtracks the original trajectory, then the algorithm is *reversible*. The Euler algorithm is not reversible, while the leapfrog algorithm is reversible.

Exercise 1.4. Time reversibility. Show numerically that the Euler algorithm is not reversible while the leapfrog is reversible. *[Hint: You can use the simulations completed in exercises 1.1 and 1.3 as a starting point.]*

Exercise 1.5. Damped harmonic oscillator. If the force also depends on the velocity, as in the presence of friction, then the total energy of the system might not be conserved, because non-conservative forces are acting on the particle. For example, consider a damped harmonic oscillator:

$$m\frac{\mathrm{d}^2 r(t)}{\mathrm{d}t^2} = -b\frac{\mathrm{d}r(t)}{\mathrm{d}t} - kr(t),$$

where $b > 0$ is the damping parameter and $-b\frac{\mathrm{d}r(t)}{\mathrm{d}t}$ is a force proportional and opposite to the velocity. Depending on the value of b, we have three different regimes, whose analytical solutions for initial conditions $r(0) = r_0$ and $v(0) = v_0$ are: the **oscillating regime**, where: $b < 2\sqrt{mk}$

$$\begin{cases} r(t) = A\, e^{-\frac{b}{2m}t} \cos(\omega t + \phi) \\ v(t) = -A\, e^{-\frac{b}{2m}t}\left[\frac{b}{2m}\cos(\omega t + \phi) + \omega \sin(\omega t + \phi)\right] \end{cases}$$

with $\omega = \sqrt{\frac{k}{m} - \left(\frac{b}{2m}\right)^2}$, $A = \sqrt{r_0^2 + \left(\frac{v_0 + \frac{b}{2m}r_0}{\omega}\right)^2}$, and ϕ defined by $\sin\phi = \frac{v_0 + \frac{b}{2m}r_0}{A\omega}$ and $\cos\phi = \frac{r_0}{A}$; the **critically damped regime**, where $b = 2\sqrt{mk}$

$$\begin{cases} r(t) = (A_1 + A_2\, t)\, e^{-\frac{b}{2m}t} \\ v(t) = \left[A_2 - \frac{b}{2m}(A_1 + A_2\, t)\right] e^{-\frac{b}{2m}t} \end{cases}$$

with $A_1 = r_0$ and $A_2 = v_0 + \frac{b}{2m}x_0$; the **overdamped regime**, where: $b > 2\sqrt{mk}$

$$\begin{cases} r(t) &= A_1\, e^{\lambda_1 t} + A_2\, e^{\lambda_2 t} \\ v(t) &= \lambda_1 A_1\, e^{\lambda_1 t} + \lambda_2 A_2\, e^{\lambda_2 t} \end{cases}$$

with $\lambda_1 = -\gamma + \sqrt{\delta}$, $\lambda_2 = -\gamma - \sqrt{\delta}$, $\gamma = \frac{b}{2m}$, $\delta = \left(\frac{b}{2m}\right)^2 - \frac{k}{m}$, $A_1 = \frac{1}{2}\left(r_0 + \frac{v_0 + \gamma r_0}{\sqrt{\delta}}\right)$, and $A_2 = \frac{1}{2}\left(r_0 - \frac{v_0 + \gamma r_0}{\sqrt{\delta}}\right)$.

a. Numerically calculate the motion of the overdamped oscillator for the three regimes using the Euler method and with the leapfrog method.

b. Compare the trajectories obtained with the two numerical methods and the analytical solutions. What are the main differences for each regime? *[Hint: compare them with the results shown in figure 1.4.]*

c. Plot the total energies of the system obtained both numerically and analytically. Check whether the work–energy theorem is numerically fulfilled, using the formulation $W_{nc} = \Delta E$, i.e., the work of the non-conservative forces is equal to the change in total energy. *[Hint: You can consider the power dissipated by the viscous force $P_{nc} = F_{nc} \cdot v = -bv^2$ and check whether it is equal to the time derivative of the total energy $\frac{dE}{dt} = krv + mva$, where r, v, a are obtained from the solution of the equation of motion.]*

1.3 Multiple particles

We now consider a system composed of multiple particles with mass m, interacting with a force derived from a pairwise potential $V(r_{ij})$, where $r_{ij} = |\mathbf{r}_i - \mathbf{r}_j|$ is the distance between the centers of two particles i and j. We will consider particles

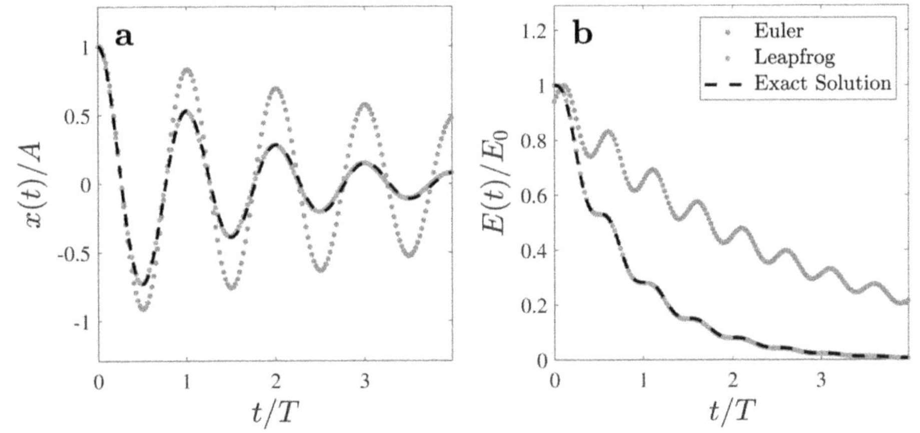

Figure 1.4. Damped harmonic oscillator. (a) Trajectory and (b) total mechanical energy of a damped harmonic oscillator ($m = 0.1$, $k = 5$, $b = 0.1\, b_{crit}$ with $b_{crit} = 2\sqrt{mk}$, $\Delta t = 0.05T$ with $T = \frac{2\pi}{\omega} = 2\pi\sqrt{\frac{m}{k}}$), which is a simple example of a system with non-conservative forces. The dashed black lines represent the analytical solution, the orange dots denote the results of a simulation with the leapfrog algorithm, and the blue dots are obtained using the Euler algorithm. The solution given by the Euler algorithm deviates significantly from the analytical solution, but it is still qualitatively reasonable, as the energy decreases exponentially with time. However, when the exponential decay of the energy is slow enough, the Euler algorithm will generate an exponential increase of the oscillation for certain time steps (which ones?).

moving in two dimensions, i.e., $\mathbf{r}_i = (r_{i,x}, r_{i,y})$ and $\mathbf{v}_i = (v_{i,x}, v_{i,y})$, but all the results can be straightforwardly generalized to higher dimensions. Building on the approach described in the previous sections, we can determine the trajectory of each particle using the system of equations:

$$\begin{cases} \dfrac{\mathrm{d}\mathbf{r}_i(t)}{\mathrm{d}t} &= \mathbf{v}_i(t) \\ \dfrac{\mathrm{d}\mathbf{v}_i(t)}{\mathrm{d}t} &= \dfrac{\mathbf{F}_i}{m} = \dfrac{1}{m}\sum_{j\neq i}\mathbf{F}_{ij} \\ \mathbf{r}_i(0) &= \mathbf{r}_{i0} \\ \mathbf{v}_i(0) &= \mathbf{v}_{i0} \end{cases} \tag{1.10}$$

where $\mathbf{F}_{ij} = \mathbf{F}(r_{ij})$ is the interaction force between particles i and j. This pairwise force permits us to describe a wide set of fundamental and effective forces, such as gravitational, electrostatic, screened electrostatic, and van der Waals forces.

The van der Waals force is particularly important for the description of real gases and can account for phenomena such as gas–liquid transitions. It is commonly modeled using the *Lennard-Jones potential* (figure 1.5):

$$V(r_{ij}) = 4\epsilon\left[\left(\frac{\sigma}{r_{ij}}\right)^{12} - \left(\frac{\sigma}{r_{ij}}\right)^{6}\right], \tag{1.11}$$

where r_{ij} is the distance between the centers of the two interacting particles, ϵ is the depth of the potential well (also referred to as the *dispersion energy*), and σ is the

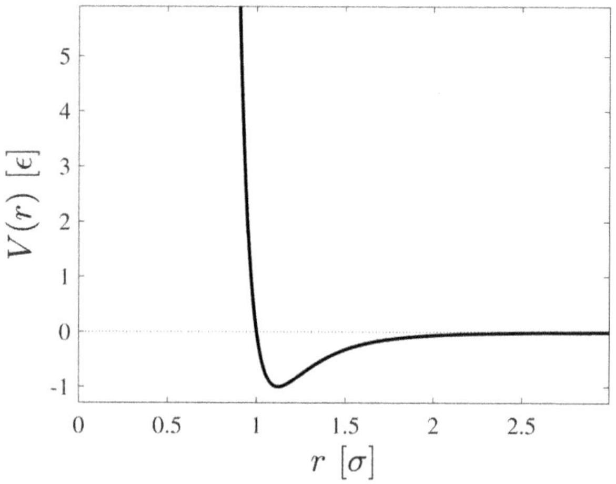

Figure 1.5. Lennard-Jones potential. The Lennard-Jones potential $V(r)$ is an effective potential that describes the essential interactions between simple atoms and molecules: two particles strongly repel each other at very small distances, weakly attract each other at intermediate distances, and do not interact at an infinite distance. Here, r is the distance between the centers of the two interacting particles, ϵ is the depth of the potential well, and σ is the distance at which the potential energy is zero. The Lennard-Jones potential reaches its minimum value at a distance $r_{\min} = 2^{1/6}\sigma$.

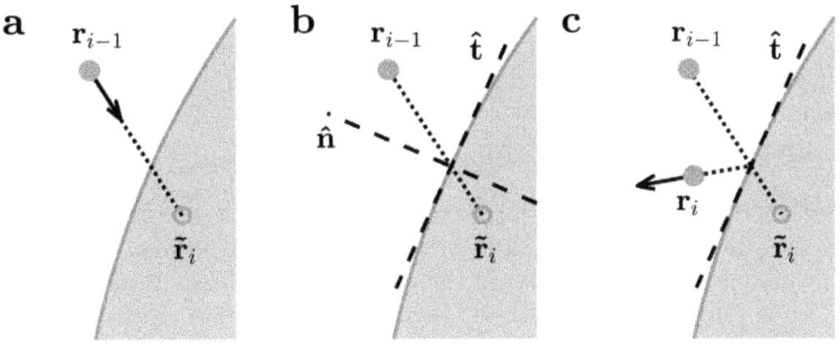

Figure 1.6. Reflecting boundary conditions. An elastic collision (i.e., a collision in which the amount of kinetic energy is conserved) of a point-like particle with a smooth, hard, fixed boundary is implemented as follows: (a) First, the intersection point of the particle trajectory with the boundary is determined. (b) The tangential and normal directions to the boundary at the intersection point are then determined. (c) Finally, the new position of the particle is determined by a mirror reflection with respect to the tangential direction. To determine the velocity of the particle at the final reflected position, we decompose the velocity of the incoming particle into its tangential and normal components, and then invert the direction of the normal component of the velocity.

distance at which the potential energy V is zero (often referred to as the *particle size*, even though it is not actually the particle size—why?).

When studying multiple interacting particles, it is convenient to confine them within a finite space. It is therefore necessary to implement the interaction of the particles with some confining boundaries. For example, we can assume that, when a particle hits the boundary of the simulation box, it bounces back without losing energy. To numerically implement this, we first define the locations of the boundaries. If, for instance, the box containing a gas is a rectangle delimited by $x = 0$ and $x = L_x$ on the x-axis and $y = 0$ and $y = L_y$ on the y-axis, whenever a particle i crosses the boundary $x = L_x$ and acquires a position $\mathbf{r}_i = (L_x + \Delta x_i,\, y_i)$ with a velocity $\mathbf{v}_i = (v_{i,x},\, v_{i,y})$, we place it back inside the container at the position $\mathbf{r}_{i,\text{refl}} = (L_x - \Delta x_i,\, y_i)$ and with the velocity $\mathbf{v}_{i,\text{refl}} = (-v_{i,x},\, v_{i,y})$, as shown in figure 1.6. We implement the reflections at the other boundaries similarly. In the reflection process, the total linear momentum of the system changes. Also, the potential energy stored in the configuration may change slightly, as the positions of some particles change. However, the kinetic energy of the system before and after the interaction with the boundary does not change, because the speed is the same before and after the reflection.

Sometimes, particles gain a lot of energy as the numerical simulation proceeds. This is an artifact, and it usually happens when the time step Δt is not small enough, even while using a time-reversible scheme, such as the leapfrog algorithm. If the time step is not small enough, two adjacent particles can come closer than should be allowed, artificially increasing the amount of total potential energy in the system.

Exercise 1.6. Two-dimensional gas in a box. Let us consider a gas of N particles (of mass $m = m_0$) interacting through van der Waals forces (equation (1.11) with $\epsilon = \epsilon_0$ and $\sigma = \sigma_0$) in a two-dimensional square container (size $L = 100\sigma_0$). Here, we are using so-called *dimensionless units* for the Lennard-Jones gas: the energy scale ϵ_0, the length scale σ_0, the mass scale m_0, the velocity scale $v_0 = \sqrt{2\epsilon/m_0}$, and the time scale $t_0 = \sigma_0\sqrt{m_0/2\epsilon}$. These dimensionless units allow us to focus our attention on the important parameters and properties of the system: for instance, if ϵ_0 is the energy scale of the interaction (i.e., the depth of the energy well of the interaction potential), providing each particle with a kinetic energy of several ϵ_0 will correspond to a high-temperature gas, while a kinetic energy of only fractions of ϵ_0 per particle can lead to the formation of molecules and/or a solid crystal/liquid phase, depending on the density of the particles. There is an additional intrinsic advantage in using dimensionless units, which is that the numbers in the simulation are close to unity, which is in the ideal precision range of most CPUs. Having numbers that are in the range of, for instance, 10^{-24} or 10^{21} (such as the values of quantities expressed in SI units) would entail some loss of precision in the numerical calculations.

a. Simulate and visualize the evolution of this gas of particles (start with $N = 100$) implementing equation (1.10) using the leapfrog algorithm. Randomize the initial positions of the particles, and orient the velocities randomly (set $v = 2v_0$ for all particles). *[Hint: Ensure that the centers of the particles are not too close to each other, because, if they are closer than σ, numerical instabilities might occur in the simulation. You can ensure this by introducing the particles one by one and checking that the center of each newly-introduced particle is not closer than σ from the centers of the other particles.]*

b. Determine an appropriate time step Δt. *[Hint: Try $\Delta t \ll \sigma/v$. In fact, if $\Delta s = v\,\Delta t$ is of the order of σ, it is easy for the simulation to fall into numerical instabilities, as the particle position changes in one time step become too large.]*

c. Check your code by plotting the instantaneous positions and velocities at each iteration for a small number of time steps. Compare your results with figure 1.7(a).

d. After you are sure your code behaves correctly, increase the number of time steps (i.e., corresponding to an interval $t_{\rm sim}$ of several t_0, ideally $t_{\rm sim} \gtrsim 100\, t_0$) and plot the kinetic, potential, and total energies of the system as a function of time. Compare your results with figure 1.7(b)–(d). *[Hint: The formulas for the kinetic and potential energy are $K = \sum_{i=1}^{N}\frac{1}{2}m(\mathbf{v}_i)^2$ and $U = \frac{1}{2}\sum_{i=1}^{N}\sum_{j\neq i}^{N} V(r_{ij})$, respectively.]*

Exercise 1.7. Three-dimensional gas. Implement and visualize a simulation of a three-dimensional gas in a box. *[Hint: Expand the simulation performed in exercise 1.6.]*

1.4 Randomness

Let us now consider the motion of a dust particle suspended in air. The air is composed of very small and fast molecules that move according to the laws of classical mechanics and that once in a while bump into the dust particle. Even though each bump is fully deterministic, the motion of the dust particle appears random. This is a simple manifestation of *Brownian motion*. It is possible to observe the emergence of Brownian motion from a deterministic molecular dynamics simulation by placing a single large particle in a gas of small particles enclosed in a box and observing the motion of the large particle, as shown in figure 1.8.

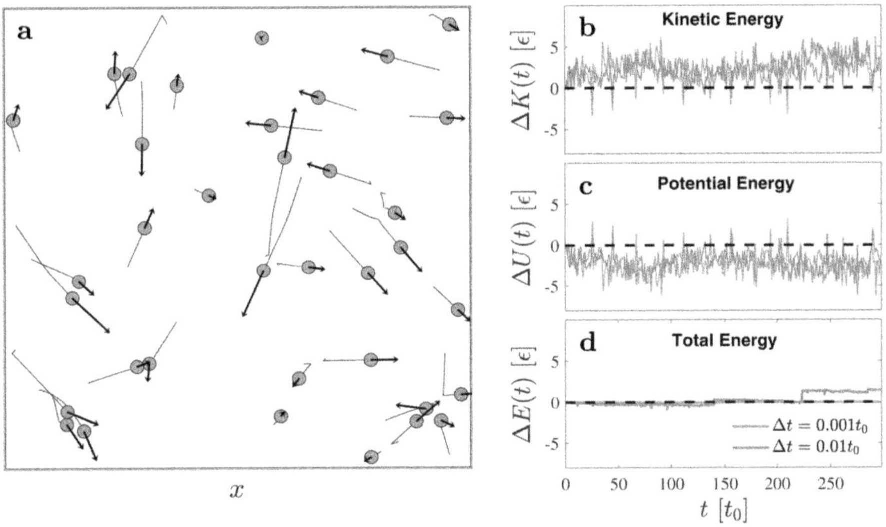

Figure 1.7. Lennard-Jones gas. (a) Typical snapshot of the instantaneous positions and velocities of gas particles (blue circles). The black arrows represent the instantaneous velocities of the particles (note that after a sufficient number of collisions, the initial conditions, according to which all particles have the same speed, are forgotten). The continuous gray lines show the last 80 time steps of the particle trajectories: the particles generally move in straight lines, except when two particles come close enough for interactions to take place. (b)–(d) Variations with respect to the initial values of the (b) kinetic, (c) potential, and (d) total energy of the Lennard-Jones gas as a function of time. The orange and blue continuous lines represent the results for $\Delta t = 0.01t_0$ and $0.001t_0$, respectively ($t_0 = \sigma_0\sqrt{m_0/2\epsilon}$; see exercise 1.6). While the kinetic and potential energy naturally fluctuate (because of the conversion from potential to kinetic energy and vice versa), the total energy is expected to remain constant. This is numerically better achieved by the simulation with the smaller time step.

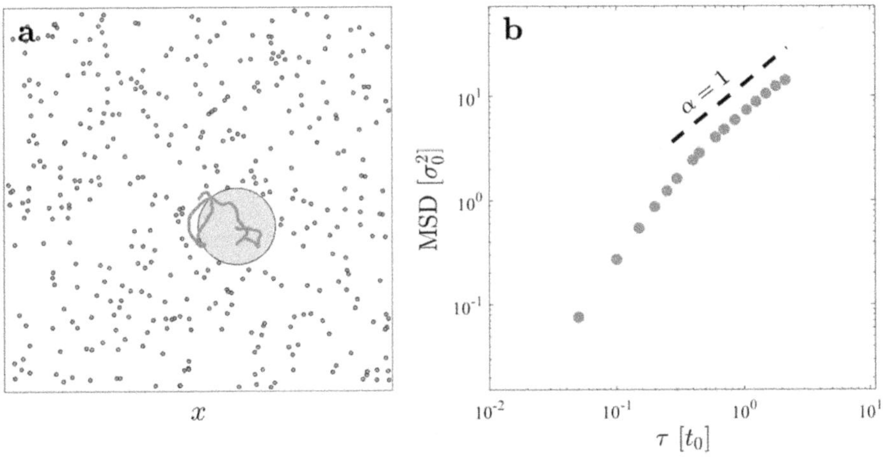

Figure 1.8. Random motion obtained from deterministic simulations. (a) A microscopic particle (the large blue circle) immersed in a gas of smaller particles (red circles) continuously undergoes collisions with the gas particles. The resulting motion obtained from a *deterministic* molecular dynamics simulation (the continuous blue line) appears to be *random*, especially if one has no access to the exact positions and velocities of the gas particles. (b) Mean square displacement (MSD) of the microscopic particle trajectory. The dashed black line emphasizes the time lags τ in which the motion of the disc is diffusive (MSD $\propto \tau$). For small time lags τ, the motion is still close to ballistic (MSD $\propto \tau^2$).

Exercise 1.8. Brownian motion. Simulate a non-interacting (for computational efficiency) gas in a box. The gas particles have mass m_0 and travel with initial speed v (refer to the dimensionless units in exercise 1.6). They are placed in the box at random positions. Add a larger particle with radius ρ and mass M that is initially at rest at the center of the box. For simplicity, the interaction between the disk and the gas particles can be modeled by a Lennard-Jones interaction using the parameters σ and ϵ, according to equation (1.11), where $r = |\,\mathbf{R} - \mathbf{r}_i\,| - \rho$, to account for the finite size of the larger particle.

a. Start your simulation with $N = 36$ gas particles, initially with speed $v = 20\, v_0$, where all the quantities are now expressed in Lennard-Jones dimensionless units. Take $\Delta t = 0.000\,05\, t_0$ (a very tiny time step, necessary given the relatively high speed of the particles) and $M = 40\, m_0$. Before starting, check that no gas particle is inside the area occupied by the larger particle. If this occurs, displace the gas particle somewhere else. Use a reflective boundary condition for the box boundaries. Run the simulation and plot its time evolution, as in figure 1.8(a).

b. Increase the number of gas particles and run again the simulation for a longer time. Save and plot the trajectory of the large particle as a function of time. Calculate the effective diffusion coefficient of the large particle. *[Hint: The diffusion coefficient D is defined using the mean square displacement (MSD) as follows:*

$$MSD(\tau) = 4\, D\, \tau.$$

Numerically, MSD(nΔt) is calculated as follows:

$$MSD(n\Delta t) = \left\langle (x_{i+n} - x_i)^2 + (y_{i+n} - y_i)^2 \right\rangle = \frac{1}{N - n} \sum_{i=1}^{N-n} (x_{i+n} - x_i)^2 + (y_{i+n} - y_i)^2.$$

Note that the MSD is not linear for all τ. If τ is small enough, the MSD is usually quadratic, because on a short time scale the collisions of the gas particles are not sufficient for the disk to lose its velocity. If the time delay is large enough, i.e., if τ is such that many collisions occur, then the information about the velocity of the disk is lost after a time lapse τ and the MSD is linear in *τ.]*

c. Change the parameters of the simulation, varying (i) the initial speed of the gas particles, (ii) the mass and (iii) the radius of the large particle. How do these changes affect the effective diffusion coefficient of the large particle?

1.5 Further reading

References [1–3] are excellent general reference books for molecular dynamics. A deep dive into their mathematical foundations and stability is provided in reference [4].

As you will have experienced in this chapter, the calculation of the force is the computational bottleneck of molecular dynamics, because its computational complexity scales according to N^2 (where N is the number of particles), while that of the rest of the code scales with N. To mitigate this problem, several methods are available that can speed up the calculations, such as building cell lists or Verlet lists [5].

Molecular dynamics lend themselves to statistical mechanics simulations. For a definition of thermodynamic macroscopic quantities in term of microscopic variables see, e.g., reference [6]. An excellent book on statistical thermodynamics, with insightful numerical exercises, is reference [7].

The methods we have seen in this chapter conserve particle number, volume, and total energy, i.e., they describe the time evolution of a system in the *microcanonical ensemble*. To simulate physical processes that happen at a constant temperature, we have to add a *thermostat* to our simulations, i.e., a step in which we add (or remove) energy to (from) our system in such a way that the temperature is kept constant. There are several ways to implement a thermostat; see, for example, references [8–11]. If we need to simulate a process at constant pressure, instead of the microcanonical ensemble, we should use the *canonical ensemble*, in which the macroscopic quantities of temperature, particle number, and pressure are conserved. To simulate a process at constant pressure, we need a *barostat*; common barostats are, for example, described in references [8–12].

Another important concept that we have not discussed in this chapter is the equilibration of the system. In fact, the initial conditions from which we start are not usually the equilibrium conditions for a thermodynamic system: it is good practice, in molecular dynamics, to let the system evolve for *enough* time before starting measuring the quantities of interest. A detailed discussion of equilibration can be found, e.g., in references [1–3].

There are also several software packages that can perform molecular dynamics simulations. One of the most used, fastest, and most efficient is LAMMPS [13], which is also conveniently open-source and free to use.

1.6 Problems

Problem 1.1. Forced oscillation. A child swinging on a swing, a vibrating guitar chord, an alternate electromotive force powering an LRC circuit, and a building subject to longitudinal earthquake waves are all examples of forced harmonic oscillation. A forced harmonic oscillator is characterized by the equation

$$m\frac{\mathrm{d}^2x}{\mathrm{d}t^2}(t) = -b\frac{\mathrm{d}x}{\mathrm{d}t}(t) - kx(t) + F(t),$$

where $b > 0$ and $F(t) = F_0\cos(\omega_f t + \phi_f)$. For $b < 2\sqrt{mk}$, numerically derive the dependence of the resonance frequency diagram of the forced harmonic oscillator. *[Hint: First, calculate the natural frequency ω_0 of the harmonic oscillation with no external forcing. Then, simulate the behavior of the system for values of ω_f in the interval $\left[\frac{1}{4}\omega_0, 4\,\omega_0\right]$, and plot the amplitude of the steady-state solution as a function of the forcing frequency ω_f.]*

Problem 1.2. Planetary orbit. Simulate the trajectory of a planet orbiting the Sun. The gravitational force between the planet (mass m) and the Sun (mass M) is:

$$F(r) = -G\frac{m\,M}{r^2},$$

where the minus sign means that the force is attractive. The force is directed along a line connecting the two celestial bodies, which can be treated as point-like particles. The gravitational potential energy is

$$U(r) = -G\frac{m\,M}{r}$$

and the total energy is $E = U(r) + \frac{1}{2}mv^2 + \frac{1}{2}MV^2$, where v is the instantaneous velocity of the planet and V the velocity of the Sun. Under the assumption that $M \gg m$ and $E < 0$, show numerically that the Sun's position is *de facto* fixed in time, and that the trajectory of the planet is an ellipse. Numerically demonstrate Kepler's second and third laws. Show that the angular momentum of the planet $\mathbf{L} = \mathbf{r} \times \mathbf{p}$, where $\mathbf{r}$ is the position of the planet relative to the Sun, is a conserved quantity. Numerically find the period of revolution of the planet's orbit.

Problem 1.3. Binary star system. Simulate the dynamics of a binary star system in which the two stars have masses m_1 and m_2. Show numerically that their motions around the common center or mass are ellipses, and that the total angular momentum and total momentum of the system are conserved.

Problem 1.4. Simple pendulum. A simple pendulum is a point-like mass suspended by a massless rope of length L. One extremity of the rope is kept fixed. The mass, attached to the other extremity, is free to oscillate under the effect of a constant and uniform gravity acceleration g. The equation defining the motion of the simple pendulum is:

$$\frac{d^2\theta}{dt^2}(t) = -\frac{g}{L}\sin\theta, \tag{1.12}$$

where θ is the angle between the vertical and the line connecting the fixed point and the suspended mass. Show numerically that, for small oscillations, the solution can be written in the form of a sinusoidal function with an angular frequency of $\omega_0 = \sqrt{\frac{g}{L}}$ and a period $T_0 = 2\pi\sqrt{\frac{L}{g}}$, independent of the motion's amplitude.

Problem 1.5. Two masses connected by an ideal spring. Simulate the dynamics of two point-like masses m_1 and m_2 connected by an ideal spring with an elastic constant k and a rest length l_0. The force that each mass experiences is $F = -k(l - l_0)$, where l is the instantaneous length of the spring. When $l > l_0$, the force between the masses is attractive, while for $l < l_0$, the force is repulsive. Simulate their trajectories, first for $l_0 = 0$ and then $l_0 > 0$. Show numerically that their total angular momentum and the total momentum of the system are invariant.

Problem 1.6. Longitudinal waves. N identical point-like particles of mass m are connected by $N - 1$ identical ideal springs with an elastic constant k and a rest length $l_0 = 0$. The masses are initially at rest at positions $x_i = 0, 1\Delta x, 2\Delta x \ldots (N-1)\Delta x$, with $\Delta x = \frac{L}{N-1}$. They are constrained to move along the x-axis only. The position of the last particle is fixed. Show numerically that, if the first particle is forced to move according to $x_1(t) = A\sin\omega t$ by an external force, the oscillation is transmitted to all the other particles, effectively inducing a one-dimensional longitudinal wave.

Problem 1.7. Transverse waves. N identical point-like particles of mass m are connected by $N - 1$ identical ideal springs with an elastic constant k and a rest length $l_0 = 0$. The masses are initially at rest at positions $x_i = 0, 1\Delta x, 2\Delta x \ldots (N-1)\Delta x$, with

$\Delta x = \frac{L}{N-1}$. They are free to move in the xy-plane. The first and last particles can move in the y direction only. Show numerically that, if the first particle is forced to move according to $y_1(t) = A \sin \omega t$ by an external force, the oscillation is transmitted to all the other particles, effectively inducing a transverse wave.

1.7 Challenges

Challenge 1.1. Solar system. Simulate the trajectory of the planets in the Solar System. Is this system stable for very long times? And what could destabilize it?

Challenge 1.2. Crystalline lattice. Simulate some atoms in a crystalline simple square lattice interacting with their neighbors via a harmonic potential. Determine what happens when a single atom is displaced from equilibrium by a small quantity Δx.

Challenge 1.3. Lennard-Jones gases, liquids, and solids. Simulate a system of N point-like particles interacting via the Lennard-Jones potential. Show that, by tuning the particle density and the kinetic energy of the particles, one can reproduce the behavior of gases, liquids, and solids.

Challenge 1.4. The Carnot engine. Simulate a Carnot engine. First simulate a gas of N particles of mass m enclosed in a rectangular container, for which one of the boundaries is free to move and the other three are fixed. The movable boundary, i.e., the *piston*, has mass M. For simplicity, start with a system of non-interacting particles, in which the interaction and heat exchange happen only through the collision with the boundary of the container. Fill the box with particles and set their initial speed according to the Maxwell distribution for a certain temperature T. Adjust the particles' initial direction by imposing an initial zero total momentum. First, numerically verify the law of perfect gases $PV = Nk_BT$. Then, implement a slow expansion or compression that simulates an isothermal and adiabatic transformation by appropriately setting the velocity of the particles after a collision with the piston for the two cases. Then, choose two temperatures T_1 and T_2 with $T_1 > T_2$ and two volumes V_1 and V_2 with $V_1 < V_2$. Start with the gas at temperature T_1 and volume V_1. Isothermally expand the gas to the volume V_2, then expand the gas adiabatically until temperature T_2 is reached; compress it isothermally until V_1 is reached and compress it adiabatically until T_1 is reached.

Challenge 1.5. Acoustic guitar. Inspired by problem 1.7, simulate a guitar chord. Find the proper values of mass density and tension that tune the chord to the different musical notes.

Challenge 1.6. Bouncing carpet. Inspired by problem 1.7, simulate a bouncing carpet. Suppose that the maximum tension that each spring can sustain is T_0. Simulate the deformation induced in the carpet by setting a spherical rigid ball of mass M with a radius r at the center of the carpet. Simulate the bouncing of the ball. Under which conditions will the carpet tear apart?

Challenge 1.7. Double pendulum. A double pendulum is made of by a pendulum with a second pendulum linked to its swinging mass. The lengths of the two massless, rigid strings are fixed and equal to L_1 and L_2, and the two hanging masses are m_1 and m_2. Simulate the dynamics of the system. Show numerically that the double pendulum exhibits chaotic motion.

Challenge 1.8. Apollo 11. On 20 July 1969, the first men landed on the Moon during the Apollo 11 space mission. Find a possible trajectory for a spacecraft to take off from Earth and land on the Moon. Get inspiration from the real trajectory followed during the Apollo 11 mission.

References

[1] Allen M P and Tildesley D J 1987 *Computer Simulation of Liquids* (Oxford: Oxford Science Publications)

[2] Frenkel D and Smit B 1987 *Understanding Molecular Simulation: From Algorithms to Applications* (New York: Academic)

[3] Rapaport D C 1996 *The Art of Molecular Dynamics Simulation* (Cambridge: Cambridge University Press)

[4] Söderlind G 2017 *Numerical Methods for Differential Equations, An Introduction to Scientific Computing* (Berlin: Springer)

[5] Verlet L 1967 Computer experiments on classical fluids. I. Thermodynamical properties of Lennard-Jones molecules *Phys. Rev.* **159** 98

[6] Kittel C 1976 *Thermal Physics* (Chichester: Wiley)

[7] Sethna J P 2006 Statistical Mechanics: Entropy, Order Parameters, and Complexity *Oxford Master Series in Physics* vol 14 (Oxford: Oxford University Press)

[8] Andersen H C 1980 Molecular dynamics simulations at constant pressure and/or temperature *J. Chem. Phys.* **72** 2384

[9] Berendsen H J C, Postma J P M, van Gunsteren W F, Di Nola A and Haak J R 1984 Molecular dynamics with coupling to an external bath *J. Chem. Phys.* **81** 3684

[10] Nosé S 1984 A unified formulation of the constant temperature molecular dynamics methods *J. Chem. Phys.* **81** 511

[11] Hoover W G 1985 Canonical dynamics: equilibrium phase-space distributions *Phys. Rev.* A **31** 1695

[12] Parrinello M and Rahman A 1980 Crystal structure and pair potentials: a molecular-dynamics study *Phys. Rev. Lett.* **45** 1196

[13] DOE labs (Sandia and LLNL) 1995 LAMMPS Molecular Dynamics Simulator https://lammps.sandia.gov/

IOP Publishing

Simulation of Complex Systems

Aykut Argun, Agnese Callegari and Giovanni Volpe

Chapter 2

Ising model

Phase transitions are very common in science and in everyday life. They often feature a *critical behavior*, which means that a small *quantitative* change to a system parameter (e.g., temperature, pressure, or concentration) can generate a dramatic *qualitative* change in the system's state. For example, a material can change its state (e.g., melting and condensation) or a metal can acquire and maintain permanent magnetic properties, which is called *ferromagnetism*. Such magnetic materials are the primary component of data storage devices, such as hard disk drives, which evolved from the older magnetic cassette

Figure 2.1. Magnetic tape. Atoms in a ferromagnetic material can keep their magnetic state for a long time, such as the iron oxide coating on a magnetic tape in a music cassette. Therefore, magnetic materials are widely used to store information. Beyond magnetic tape, other examples include floppy disks, the Video Home System (VHS), digital versatile discs (DVDs), and hard disk drives (HDDs). Data can be written locally on these media using an electromagnetic writing element, which modifies the magnetic state of a domain. These magnetic domains can remain unaltered for years with no power required. Source: Cassette tape audio analog by wongpear on Pixabay under the Simplified Pixabay Licence.

doi:10.1088/978-0-7503-3843-1ch2

tapes (figure 2.1). The data written in a magnetic material can be stored unaltered for years due to the ferromagnetic properties of the material used. Ferromagnetic materials, such as iron, cobalt, and nickel, are very important for modern technology, as they are also key elements in electric motors, microphones, and speakers.

The modeling of phase transitions can be rather complex. In particular, it is necessary to understand how the short-range interactions between atoms or molecules in an atomic lattice give rise to long-range changes. In 1925, Ersnt Ising, a young scientist, came up with a simple but very powerful method to model ferromagnetism [1]. The resulting *Ising model* is now a classical tool used to understand magnetization and phase transitions. It is an example of a *Monte Carlo* simulation method.

In principle, it is possible to simulate a system directly by converting its underlying differential equations into difference equations, as we have done in chapter 1, where we used Newton's laws to implement molecular dynamics simulations. However, this is not always possible in practice because, as the size of the system grows, the ensuing set of equations becomes unmanageably computationally complex. Furthermore, we are often not even interested in the detailed trajectories of each and every atom of the system, but we would like to observe global equilibrium properties and long-range behaviors. In these cases, the Monte Carlo method becomes extremely useful. This method reconstructs the evolution of a system by sampling its subsequent states with a probability that depends on their energy. It is a very popular and powerful tool used to simulate interactions among a large number of particles, such as those of disordered materials, strongly coupled solids, spins, liquids, and cellular structures [2].

In this chapter, we start with the fundamentals of the Monte Carlo method and how it is used in statistical physics in conjunction with the *Boltzmann distribution*. Next, we apply the Monte Carlo method to simulate a two-dimensional lattice of spins that features ferromagnetic properties (i.e., the Ising model). In particular, we observe how the state of the material is affected by the temperature and by external magnetic fields. Finally, we simulate *critical demixing* and the *critical Casimir forces* that can arise between particles in its presence.

Example codes: Example Python scripts related to this chapter can be found at: https://github.com/softmatterlab/SOCS/tree/main/Chapter_02_Ising_Model. Readers are welcome to participate in the discussions related to this chapter at: https://github.com/softmatterlab/SOCS/discussions/8.

2.1 Monte Carlo method

The Monte Carlo method includes a wide class of numerical algorithms that relies on the random sampling of the phase space of a system. Its common applications are optimization problems, numerical integration, and, in particular, in statistical mechanics, the simulation of large ensembles of particles in order to determine their physical properties. At each iteration, the states of the particles are updated according to a probability distribution that depends on the energy levels of the possible states and on the temperature of the environment.

For the case of a single particle, the Monte Carlo method works as follows: given a particle that can assume a certain number of states indexed by i with energy E_i, the Monte Carlo method changes the state of this particle according to the probability distribution

$$p_i = \frac{1}{Z} e^{-\frac{E_i}{k_B T}}, \tag{2.1}$$

where $Z = \sum_i e^{-\frac{E_i}{k_B T}}$ is a normalization factor (the so-called *partition function*), k_B is the Boltzmann constant, and T is the temperature of the environment. The simplicity of the Monte Carlo method relies on the use of randomness to determine to the correct equilibrium state of a system through stochastic transitions. However, it is important to keep in mind that the Monte Carlo method determines the equilibrium properties of a system, but it does not necessarily reproduce the correct dynamics of the same system.

Exercise 2.1. Kramers transitions. Let us consider a system with two identical stable states (*ground states*) separated by an *energy barrier* E_b. This system can easily be modeled with a one-dimensional three-state system, in which a particle can only transition into the neighboring state:

Left $\Longleftrightarrow$ Middle $\Longleftrightarrow$ Right

where the 'Middle' state corresponds to the energy barrier with energy E_b, and the 'Left' and 'Right' states have zero energy. At each step, the possible future states are only the present state and its neighboring states (e.g., a transition from 'Left' to 'Right' is not possible). This model is a simplified version of the very powerful model introduced in 1940 by Hans Kramers to estimate the frequencies of chemical transitions [3].

a. Simulate the system setting $E_b = 2k_B T$, starting at the 'Left' state and running the simulation for 10^5 steps. Check the trajectories of the system and the equilibrium probability distributions. How many steps are necessary to reach the equilibrium distribution?

b. Vary E_b and T to observe their effect on the transition frequency between states and on the equilibrium distribution. Does this affect also the time it takes the system to reach the equilibrium distribution?

c. Start the simulation in the 'Left' state and stop it whenever it reaches the 'Right' state, recording the corresponding *escape time*. Repeat this numerical experiment many times in order to determine the escape time distribution for different values of E_b and T.

2.2 Ising model

The Ising model considers a two-dimensional square lattice of atoms, each with a spin that can have two states: atomic spin up, or atomic spin down. Each atom interacts with its neighboring atoms and with an external magnetic field, as shown in Fig 2.2(a). The energy of the atom at position (i, j) has the following expression [4]:

$$E_{i,j} = -H\sigma_{i,j} - J\sigma_{i,j} \sum \sigma_{i\pm 1, j\pm 1}, \tag{2.2}$$

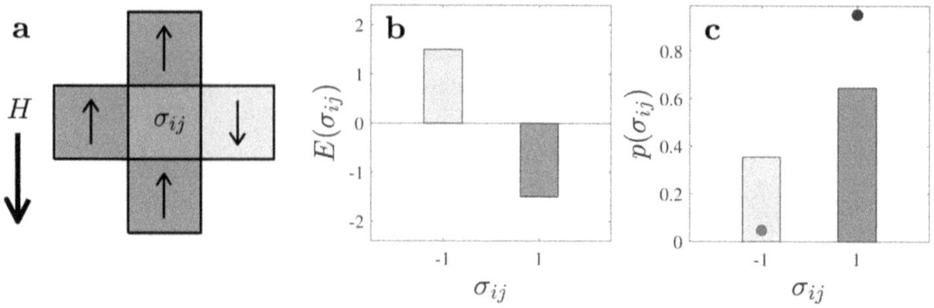

Figure 2.2. Ising model. (a) Each individual atom in the lattice has two accessible magnetic states that are -1 (downward arrow, yellow lattice site) and $+1$ (upward arrow, blue lattice sites). At every step, we randomly change one of the spins (σ_{ij}, orange lattice site) with a probability that depends on its neighbors' spins and on the external magnetic field H. (b) The energies of the states $\sigma_{ij} = \pm 1$ are calculated using equation (2.2) with $J = 1$ (ferromagnetic material) and $H = -0.5$. (c) The corresponding probability distributions are calculated according to equation (2.1) with $T = 5$ ($k_B = 1$). Note that even though the energy values are quite different, the probability of choosing each state is still comparable when the temperature is high. If the temperature was significantly lower, these probabilities would be a lot more asymmetrical (as shown by darker dots for $T = 1$).

where $\sigma_{i,j}$ is the spin of the atom at position (i, j), J is the spin–spin coupling strength, and H is the intensity of the external magnetic field. Without loss of generality, we can set $J = 1$ and $k_B = 1$. It is worth noting that this model reduces the complexity of spin coupling in the lattice down to one parameter, namely the sign of J. If J is positive, the ensuing behavior is ferromagnetic [5]. If J is negative, this causes the neighboring spins to align oppositely to each other in order to minimize the energy, resulting in antiferromagnetic behavior.

We now list the steps required to simulate the two-dimensional Ising model using the Monte Carlo method:

1. Initialize a lattice of size $N \times N$ with random initial spins $\sigma = \pm 1$.
2. Select a random atom at position (i, j).
3. Calculate the sum of the four neighboring spins: $M = \sigma_{i-1,j} + \sigma_{i+1,j} + \sigma_{i,j-1} + \sigma_{i,j+1}$, as shown in figure 2.2(a).
4. Calculate the energies of the possible future states: $E_+ = -(H + JM)$ (spin up) and $E_- = H + JM$ (spin down), as shown in figure 2.2(b).
5. Apply the Monte Carlo algorithm (equation (2.1)) to recalculate this spin. In other words, set the spin of this atom to $+1$ (up) with a probability of $\frac{e^{-\beta E_+}}{e^{-\beta E_+} + e^{-\beta E_-}}$ and to -1 (down) with a probability of $\frac{e^{-\beta E_-}}{e^{-\beta E_+} + e^{-\beta E_-}}$, as shown in figure 2.2(c).

Temperature is the most important parameter in the Ising model (figure 2.3), as we will discover in this chapter. As the temperature becomes low (e.g., $T = 1$, keeping in mind that we have set $k_B = 1$ so that $\beta = T^{-1}$), the individual spins become much more likely to adopt the same magnetization as their neighbors. This results in the formation of large domains with the same magnetization. Once these large magnetic domains are formed, individual spin fluctuations with the opposite magnetization become highly unlikely inside a domain; therefore, such a material can preserve its magnetic properties

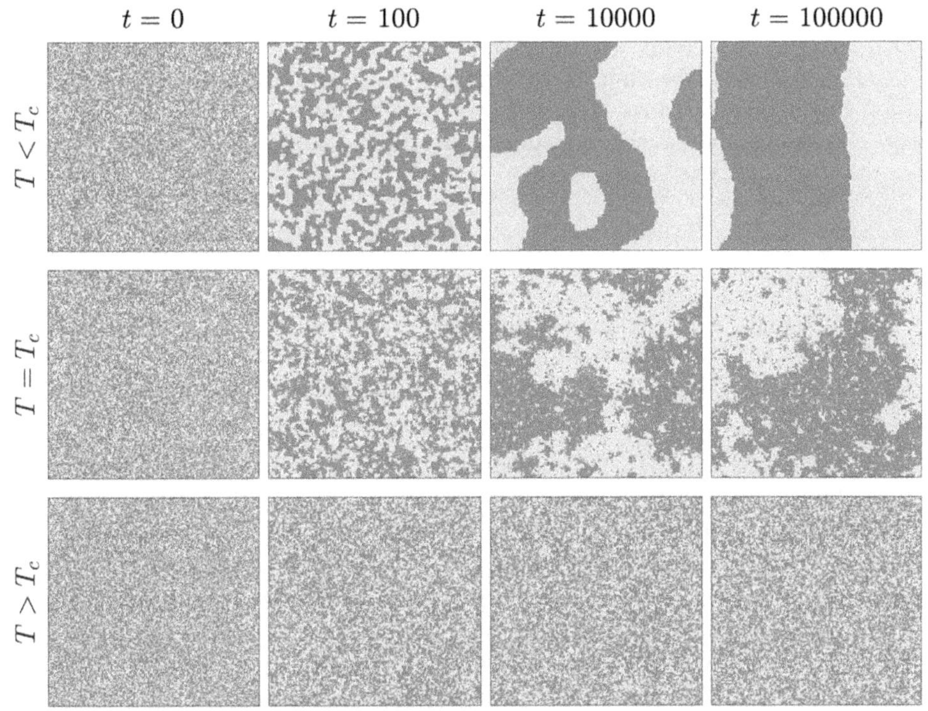

Figure 2.3. Snapshots of a 2D Ising model. Spin map snapshots of a 2D Ising model after 0, 100, 100 00, and 100 000 steps at various temperatures T ($H = 0$). Below the critical temperature (top row, $T = 1 < T_c$), the magnetic states rapidly converge into permanent magnetic domains that have spins of either +1 or −1. At the critical temperature (middle row, $T = T_c = 2.269$), domains of opposite magnetization form, although fluctuations are prominent. Above the critical temperature (bottom row, $T > T_c$), fluctuations are dominant and the random texture is preserved over time; large magnetized domains do not form. The results are shown for a lattice of 200 × 200 atoms ($J = 1$, $k_B = 1$) updating 10% of the spins at each step.

for a very long time. However, if the temperature is high (e.g., $T = 5$), the probabilities of different spins are a lot closer, despite the difference in energy levels; thus, stable domains never form at high temperatures, as fluctuations are dominant.

A standard measure of the state of a magnetic material is its magnetization per unit volume:

$$m = \frac{1}{N^2} \sum_{i,j} \sigma_{i,j}. \tag{2.3}$$

This can also be used to measure a material's response to external magnetic fields.

Exercise 2.2. Ising model on a 2D lattice. Write a program that simulates an Ising model with 200 × 200 cells. Initialize the spins randomly with an equal probability of being set to +1 or −1. At each step, randomly pick 10% of the atoms in the lattice and update their states by following the Monte Carlo method.

a. Plot some snapshots of the binary spin map of the material. Repeat the same numerical experiment for various temperatures T while the external magnetic field is absent ($H = 0$). Observe the clustering of certain magnetic domains at low temperatures and the more random texture of the lattice at higher temperatures. *[Hint: See figure 2.3.]*

b. This system has a critical temperature of $T_c = 2.269$. Simulate the different behaviors at subcritical and supercritical temperatures and critically analyze the differences between them. *[Hint: See first and last row in figure 2.3.]*

c. Show that the 2D Ising model behaves like a paramagnetic material at higher temperatures (e.g., $T = 5$).

d. Calculate the total magnetization of this system as a function of H. Show that, for small values of H, the magnetization density of the system scales according to $m = \chi H$, where χ is the *magnetic susceptibility*. Calculate χ.

As we have seen, at high temperatures, the structure of the 2D Ising model is random. However, this system can respond to an external field H, as we have seen in the previous exercise. By contrast, when the system is at low temperatures, the system becomes magnetized even if there is no external magnetic field. This is due to the fact that neighboring spins affect each other very strongly and the magnetic domains grow.

If the interaction between spins is *antiferromagnetic* (i.e., $J < 0$), neighboring atoms tend to align oppositely to each other. While such materials show a similar (disordered) behavior to those of ferromagnetic ones at high temperatures, they strongly resist magnetization under the influence of an external field, even at low temperatures.

Exercise 2.3. Antiferromagnetic materials. Write a program that simulates an Ising model with 200 × 200 cells that is antiferromagnetic ($J = -1$).

a. Show that, for high temperatures, the magnetic susceptibility coefficient of an antiferromagnetic material is very similar to that of a ferromagnetic one.

b. Show that, at low temperatures, unlike ferromagnetic materials, antiferromagnetic materials does not magnetize. Plot the average magnetization m as a function of the temperature and the external magnetic field.

2.3 Critical temperature

A ferromagnetic material becomes magnetized when the temperature is below a *critical temperature* T_c (also known as the *Curie temperature*). Most natural ferromagnetic materials have a Curie temperature that is much higher than room temperature; therefore, they are commonly found in their magnetic state in everyday life [5].

If the temperature is very low (i.e., much lower than the critical temperature), most spins acquire the same orientation, leading to a strong average magnetization (even in the absence of external magnetic fields). This is demonstrated in figure 2.4(a), in which a randomly-initialized system often reaches a maximum magnetization with spins of either +1 or −1. Instead, if the temperature is very high

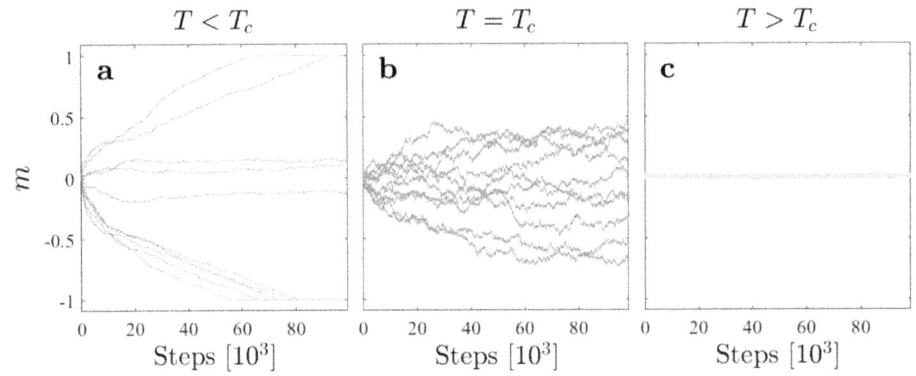

Figure 2.4. Divergence of the average magnetization for a 2D Ising model. Study of average magnetization with ten independent simulation runs of the 2D Ising model for different temperatures ($J = 1$, $H = 0$, $N = 200$, $k_B = 1$). (a) When the temperature of the system is low ($T = 1$), the system is unstable and quickly polarizes randomly into big domains of the same magnetization, even when there is no external magnetic field. (b) When the temperature is equal to the critical temperature ($T = T_c = 2.269$), the average magnetization oscillates slowly over time. (c) When the temperature of the system is high ($T = 5$), the average magnetization fluctuates around zero.

(i.e., much higher than the critical temperature), the orientation of a spin is largely independent of those of its neighbors, resulting in an uncorrelated random spin pattern and a zero average magnetization. This is shown in figure 2.4(c), where the magnetization only fluctuates around zero. At temperatures close to the critical temperature, *criticality* emerges: large domains form with one particular spin, although the overall average magnetization is still zero, leading to slow oscillations of the overall magnetization of the system (figure 2.4(b)). The Ising model reproduces this whole range of behaviors.

Exercise 2.4. Criticality. Write a program to simulate a 200×200 Ising model and explore its critical behavior. Choose three temperature values (e.g., $T = 1$, $T = 2.269$ and $T = 5$) and run the simulation for 10^5 steps, updating 10% of the spins in each step.

- **a.** Calculate the magnetization as a function of the step number for all three temperatures and repeat each simulation ten times in order to observe the statistical behavior. Plot the results as shown in figure 2.4.
- **b.** Show that, for $T < T_c$, the magnetization eventually diverges away from zero in either direction—the faster the divergence, the lower the temperature (blue lines in figure 2.4(a)). For $T > T_c$, the magnetization eventually only fluctuates around zero (yellow lines in figure 2.4(c)).
- **c.** Calculate the variance of the magnetization of different simulation runs as a function of the step number. Show that this variance saturates to a steady-state value for temperatures lower and higher than the critical temperature $T_c = 2.269$, but not at T_c.

Exercise 2.5. Measuring the critical temperature. The Ising model is computationally heavy, especially near the critical temperature, because the divergence of the magnetization happens extremely slowly. Repeat the simulation for a wide range of

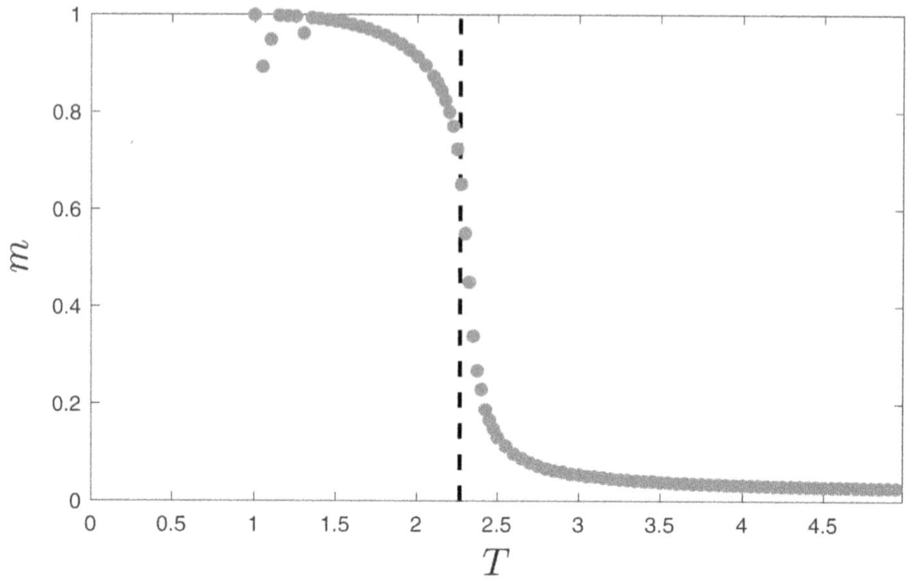

Figure 2.5. Critical temperature of the 2D Ising model. The average absolute magnetization m (blue dots, equation (2.3)) as a function of temperature. The critical temperature is $T_c = 2.269$ (black dashed line). Each data point represents the magnetization of the lattice after 10^6 steps averaged over ten simulation runs. The material is ferromagnetic ($J = 1$) and there is no external magnetic field ($H = 0$).

temperatures ($H = 0$) and obtain the average magnetization for each temperature, as shown in figure 2.5. Show that the critical temperature is $T_c = 2.269$.

Exercise 2.6. Domain size and correlation length. Quantify the correlation length of the spins in the 2D Ising model and the size of the domains. How are these two quantities related? How do they depend on temperature and external magnetic fields?

2.4 Critical mixtures

The Ising model can also be employed to model critical liquid mixtures. A critical liquid mixture is formed by two liquids that dissolve homogeneously within each other at temperatures above a critical temperature and demix below the critical temperature. Upon approaching the critical temperature from above, the thermal fluctuations induce a local variation of the concentration of the mixture components: the spatial correlation length ξ of these concentration fluctuations depends on the temperature difference $\Delta T = T - T_c$ via a critical universal exponent $\nu = 0.63$:

$$\xi \propto \left(\frac{\Delta T}{T_c}\right)^{-\nu}. \tag{2.4}$$

Hence, the correlation length of the concentration fluctuations increases with no limit for $T \rightarrow T_c$. The critical fluctuations give rise to the phenomenon of *critical opalescence*: when the temperature is sufficiently close to T_c, the solution becomes optically inhomogeneous, and changes from completely transparent to turbid.

When the critical fluctuations are *spatially confined* by particles suspended in the solution, they might induce *forces* between the particles—the so-called *critical Casimir forces*. These forces become effective when the correlation length of the concentration fluctuations is comparable to the separation between the particles. They can be attractive or repulsive, mainly depending on the preferential adsorption properties of the surfaces confining the fluctuations. Usually, so-called *symmetric boundary conditions* (i.e., when both particles are attracted to the same component of the mixture) induce attraction, while *antisymmetric boundary conditions* (i.e., when the two particles are attracted to different components of the mixture) induce repulsion. In the next exercise, we will simulate the critical Casimir forces using an effective model.

Exercise 2.7. Critical Casimir forces. Simulate a 100×100 2D Ising model at a temperature slightly above the critical temperature.

- **a.** Place two square particles of size 20×20 within the simulation arena at a fixed position (see, e.g., figure 2.6(a)). Fix the value of the spins within these particles to $+1$. This means that the surface properties of both particles are the same, i.e., both particles have a stronger affinity for $+1$ spins than for -1 spins. Run the simulation under these conditions for a long time and measure the total energy of the system.
- **b.** Calculate the energy of the system for different distances between particles and plot it as a function of distance. Use this information to calculate the critical Casimir forces between the particles. *[Hint: The force is the minus derivative of the potential.]*
- **c.** Explore what happens when one particle has a spin of $+1$ and the other has a spin of -1, so that the two particles have surface affinities for opposite spins (see, e.g., figure 2.6(b)). *[Hint: We expect a repulsive force between the particles. Why?]*

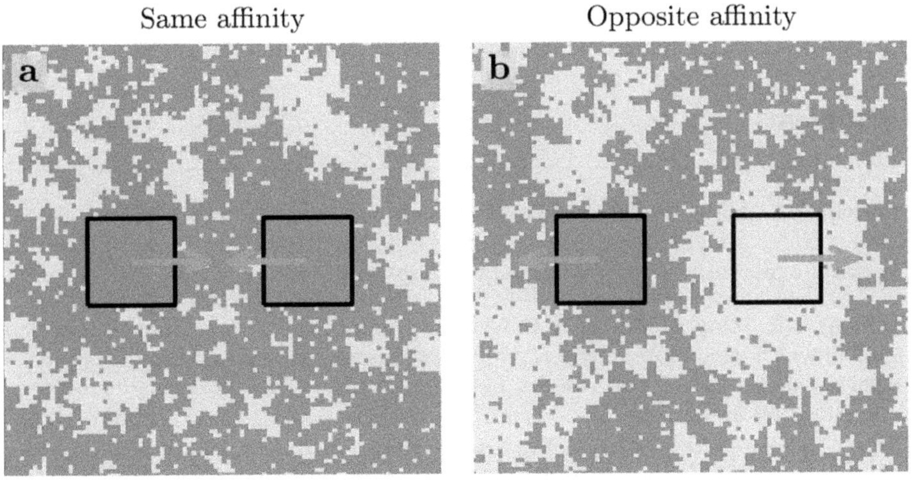

Figure 2.6. Critical Casimir forces generated by critical fluctuations. Snapshot of an Ising model at a temperature slightly above the critical temperature with two particles (a) with the same affinity and (b) with opposite affinities. Because of the critical fluctuations, attractive (repulsive) forces arise between particles with the same (opposite) affinity, as indicated by the arrows.

An example of a critical binary mixture is water–lutidine at the critical concentration. Water consists of polar molecules and lutidine is an oil-like substance. This particular system has a second-order *lower* critical point (the 2D Ising model has an *upper* critical point instead) at room temperature at about $T_c = 34°C$: at temperatures below T_c, the mixture is homogeneous; at temperatures above T_c, it demixes into its water and lutidine phase.

2.5 Further reading

The original Ising model was formulated in one dimension, in which no phase transition and consequently no critical exponent were found [1]. The 2D Ising model has an exact solution, found by Lars Onsager [6]. An exact solution for the 3D Ising model has not yet been found, and it is believed not to exist, but there is no proof of this.

Excellent general reference books for Monte Carlo algorithms and the Ising model are references [2, 4]. A comprehensive theoretical study with a large number of case studies is provided in reference [7].

The Ising model is computationally heavy, as the fluctuations near the critical temperature require a very large number of simulation steps to be quantified. Fortunately, the Ising model with the Monte Carlo method is easily parallelizable [8] (e.g., with the use of GPU computing, the speed of these simulations can be accelerated by hundreds of times [9]).

The Monte Carlo method has been widely used in many different areas of the computational sciences [10], for example, for industrial optimization and reliability [11], for a wide range of physical systems from particle physics to astronomy [12–15], and for economics and finance [16, 17].

Critical Casimir forces have been measured experimentally [18]. They have been predicted to be *non-additive* [19], and non-additivity for spherical particles in the bulk has been measured experimentally [20]. Critical Casimir forces are important for the controlled and reversible self-assembly of colloids [21].

2.6 Problems

Problem 2.1. Ising model in 1D. Write a program for a one-dimensional Ising model in which each atom interacts with its two nearest neighbors. Show that there are no critical temperatures and no phase transitions in this case.

Problem 2.2. Ising model in 3D. Write a program for a three-dimensional Ising model in which each atom interacts with its six nearest neighbors. Study the model as we have done for the 2D Ising model. In particular, determine the critical temperature of the 3D Ising model.

Problem 2.3. Magnetic coercivity. Ferromagnetic materials show resistance to an external magnetic field if they have been polarized in the opposite direction. Let us analyze this effect.

a. Consider an Ising model at a temperature below the critical temperature (200×200, $J = 1$, $k_B = 1$, $T = 0.5$). Initially, set all the spins to downward ($\sigma_{ij} = -1$ for all

i, j). Apply an external field H and observe that for small values of $H > 0$, the material does not flip its magnetization to align itself with the external field. Find the value of H for which the magnetization flips.

b. Repeat for several values of T and plot the necessary external field H that flips the magnetization as a function of T.

Problem 2.4. Interplay between ferromagnetic and antiferromagnetic behavior. Study the behavior of an ensemble of atoms of types A and B that forms a square lattice such that each A has four side neighbors that are B and four diagonal neighbors that are A, and vice versa for B. Assume that the coupling strengths between atoms are given by J_{AA}, J_{AB} and J_{BB}. If $J_{AB} = 0$, this system will have two different critical temperatures according to J_{AA} and J_{BB}.

a. Start with no cross-type interactions and observe the different critical temperatures of the lattice. Then, introduce a nonzero J_{AB} and analyze what happens to the critical temperature values of the system.

b. Make one of the couplings antiferromagnetic ($J_{BB} < 0$). Observe what happens as you increase the cross-type coupling J_{AB}.

Problem 2.5. Non-additivity of critical Casimir forces. Study the non-additivity of critical Casimir forces. In fact, the critical Casimir forces that we measured in exercise 2.7 are non-additive.

a. Start by making a 100×100 Ising model simulation that has two disk particles with the same affinity. Follow the same procedure as for exercise 2.7 and determine the forces between the two disks.

b. Place three disks at equal distances from each other and calculate the forces acting on them. Show that the resulting total force is not equal to the force obtained by adding up the pairwise interactions between the three pairs of particles.

c. Study the effects of different particle affinities.

2.7 Challenges

Challenge 2.1. Ising model on a graph. Simulate the Ising model on a graph, starting with a regular graph such as a Kagome lattice (an equilateral triangular lattice) or on an hexagonal lattice, and then move toward more complex graphs (check chapter 12 for inspiration about what kinds of graph to consider).

Challenge 2.2. Continuous spins. Generalize the binary Ising model into a continuous model in which each spin can be oriented in any random direction $\theta \in [-\pi, \pi]$. Implement the updated Monte Carlo method with continuous energy levels selected randomly at each update. Calculate the critical temperature of this system.

Challenge 2.3. Magnetic data storage on a cassette tape. Realize an Ising model on a long magnetic tape of 50×5000 lattice sites. Assume that an external magnetic field of the form $H(x, t) = H(t, x - vt)$ is applied to the tape in order to store information. Find out which parameters of the model (J, T, H) would make it possible to reliably write, read, erase, and rewrite data. Design your own function $H(x, t)$.

References

[1] Ising E 1925 Beitrag zur Theorie des Ferromagnetismus *Z. Phys.* **31** 253

[2] Newman M and Barkema G 1999 *Monte Carlo Methods in Statistical Physics* vol 24 (Oxford: Oxford University Press)

[3] Kramers H A 1940 Brownian motion in a field of force and the diffusion model of chemical reactions *Physica* **7** 284

[4] Sethna J P 2006 Statistical Mechanics: Entropy, Order Parameters, and Complexity *Oxford Master Series in Physics* vol 14 (Oxford: Oxford University Press)

[5] Kittel C and McEuen P 1976 *Introduction to Solid State Physics* 8th edn (New York: Wiley)

[6] Onsager L 1944 I. A two-dimensional model with an order-disorder transition *Phys. Rev., Series II* **65** 117

[7] Newell G F and Montroll E W 1953 On the theory of the Ising model of ferromagnetism *Rev. Mod. Phys.* **25** 353

[8] Maigne L *et al* 2004 Parallelization of Monte Carlo simulations and submission to a grid environment *Parallel Process. Lett.* **14** 177

[9] Preis T, Virnau P, Paul W and Schneider J J 2009 GPU accelerated Monte Carlo simulation of the 2D and 3D Ising model *J. Comput. Phys.* **228** 4468

[10] Gentle J E 2006 *Random Number Generation and Monte Carlo Methods* (Berlin: Springer)

[11] Elperin T, Gertsbakh I and Lomonosov M 1991 Estimation of network reliability using graph evolution models *IEEE Trans. Reliab.* **40** 572

[12] Metropolis N, Rosenbluth A W, Rosenbluth M N, Teller A H and Teller E 1953 equation of state calculations by fast computing machines *J. Chem. Phys.* **21** 1087

[13] Sauvan P, Sanz J and Ogando F 2010 New capabilities for Monte Carlo simulation of deuteron transport and secondary products generation *Nucl. Instrum. Methods. Phys. Res.* A **614** 323

[14] Gillespie D T 1976 A general method for numerically simulating the stochastic time evolution of coupled chemical reactions *J. Comput. Phys.* **22** 403

[15] Springel V 2005 The cosmological simulation code GADGET-2 *Mon. Not. R. Astron. Soc.* **364** 1105

[16] Boyle P P 1977 Options: a Monte Carlo approach *J. Financ. Econ.* **4** 323

[17] Giles M B 2008 Multilevel Monte Carlo path simulation *Oper. Res.* **56** 607

[18] Hertlein C, Helden L, Gambassi A, Dietrich S and Bechinger C 2008 Direct measurement of critical Casimir forces *Nature* **451** 172

[19] Mattos T G, Harnau L and Dietrich S 2013 Many-body effects for critical Casimir forces *J. Chem. Phys.* **138** 074704

[20] Paladugu S *et al* 2016 Nonadditivity of critical Casimir forces *Nat. Commun.* **7** 11403

[21] Nguyen T A, Newton A, Kraft D J, Bolhuis P G and Schall P 2017 Tuning patchy bonds induced by critical Casimir forces *Materials* **10** 1265

IOP Publishing

Simulation of Complex Systems

Aykut Argun, Agnese Callegari and Giovanni Volpe

Chapter 3

Forest fires

Landslides, earthquakes, floods, and wildfires are examples of natural disasters that can have a high environmental, societal, and economic cost. Civil protection programs aim to minimize the human and material losses that originate from natural disasters, and their deployment requires a careful risk assessment. Such risk assessment is based both on models of the dynamics of the hazard in question and on historical analysis of its past occurrences. Interestingly, the laws describing such natural phenomena are often *scale invariant*, i.e., they have a universal character that does not change with changes to the scales of time, space, energy, or other relevant quantities. This is often reflected by the fact that the properties of the system (e.g., the size or frequency of events) follow a *power law*. Everyday examples of scale invariance are the branches of a tree or the structure of a Romanesco broccoli, as shown in figure 3.1. In fact, these are natural examples of *fractals*, which are objects whose parts have the same statistical properties as the whole. Self-similarity is also present in critical systems near a phase transition, such as, for example, the Ising model near its critical point in chapter 2.

Figure 3.1. Scale invariance in a Romanesco broccoli. The structure of a Romanesco broccoli is a beautiful example of scale invariance: if we take a small part of the broccoli (left panel, white frame) and we appropriately rescale and rotate it (central panel), we obtain a replica of the whole broccoli. The process can be repeated multiple times (right panel). *Source: adapted from Romanesco broccoli (Brassica oleracea), Wikipedia: Ivar Leidus (CC BY-SA 4.0).*

doi:10.1088/978-0-7503-3843-1ch3

Sometimes, the universal behavior observed in scale-invariant systems does not depend on finely tuned parameters. Instead, the complex behavior is robust and emerges even if the system parameters vary widely. There are cases in which the mechanism by which complexity emerges is *spontaneous*, thus opening the possibility of naturally-occurring complexity, without the need to fine-tune the system near very special conditions. In the literature, this concept has been referred to as *self-organized criticality* [1, 2]. The essential idea behind self-organized criticality is that something intrinsic to the system drives it toward a critical point, regardless of the initial conditions. Large-scale natural phenomena displaying scale-invariant behavior, such as earthquakes, solar flares, landscape formation, forest fires, and landslides, have been modeled and analyzed from the perspective of self-organized criticality.

In this chapter, we present a simplified scale-invariant model for forest fires. We study forests exposed to periodic fires, showing that they feature a self-organized critical behavior. We start by setting the rules for forest growth, the frequency of ignition of a tree, and the consequent propagation of the fire, which burns down part of the forest. We then study the statistical properties of the resulting dynamical system. In particular, we analyze the size distribution of the forest fires, which follows a power law that is characteristic of critical systems. The system under examination is thus an example of self-organized criticality.

Example codes: Example Python scripts related to this chapter can be found at: https://github.com/softmatterlab/SOCS/tree/main/Chapter_03_Forest_Fires. Readers are welcome to participate in the discussions related to this chapter at: https://github.com/softmatterlab/SOCS/discussions/10.

3.1 Forest growth and fire ignition

We model the area where the trees grow as a square lattice with $N \times N$ cells, as shown in figure 3.2(a). Each cell can be empty or occupied by one tree. At the beginning (time $t = 0$), we start with an empty forest. At each time step, there is a probability p that a tree grows in each of the empty sites. All new trees grow independently from each other, and their growth does not depend on their proximity to other trees. At each time step, there is a probability f that a single lightning bolt strikes the forest at a random position. Each cell has the same probability of being hit: if the hit cell is empty, no fire is started. If the hit cell contains a tree, the tree is ignited and a forest fire starts, as shown in figure 3.2(a).

The timescale of a forest fire is much faster that the timescale of tree growth: when a fire occurs, it burns down everything it can. Therefore, we assume that when a tree is ignited, the fire propagates instantaneously (i.e., within the same time step) to all trees that belong to the same cluster [3]. We consider that two trees belong to the same cluster if there is at least one connecting path passing through the neighboring sites that are each occupied by a tree, as shown in figure 3.2(c). To start with, we consider the *von Neumann neighborhood* (i.e., a cell is the first neighbor of the four cells with which it shares a side (right, left, top, and bottom), as shown in figure 3.2(b)). To determine the

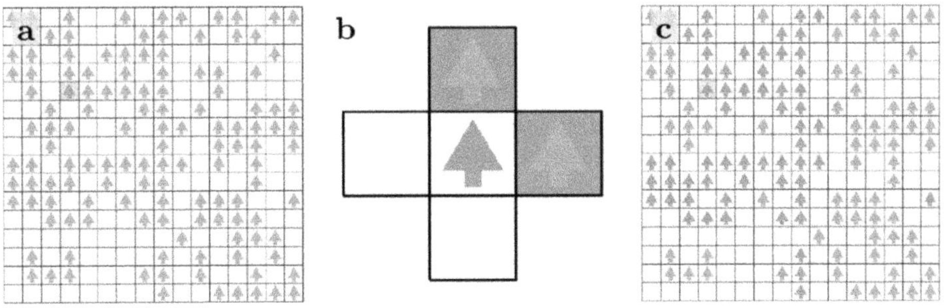

Figure 3.2. Rules for forest growth and tree ignition. (a) A forest with 16×16 cells; the trees are indicated by green pines. At each time step, there is a probability p that a new tree grows in each empty cell and a probability f that a lightning bolt hits a random cell (indicated by the yellow cell), which starts a fire if the hit cell is occupied by a tree. (b) Each fire spreads to all the von Neumann neighbors (top, bottom, left, and right cells) of every burning tree, if they are occupied by another tree (orange cells, top and right). (c) When a lightning bolt strikes (the cell with the yellow background) a cell with a tree, the tree catches fire. The fire then propagates to the cluster of connected trees (orange pines). In this case, we have considered periodic boundary conditions, as one can see from the apparently isolated burning tree on the right side: its neighbor is the burning tree at the other end of the same row.

members of the cluster, we must keep in mind that the chosen (periodic or closed) boundary conditions play a role; in the case of periodic boundary conditions (which we consider in this chapter), we have to consider the proximity of trees on the other side of the square domain, and we have to propagate the fire accordingly. For example, in figure 3.2(c), the apparently isolated burning tree on the right-hand side is, in reality, a first neighbor of the burning tree located in the same row, but on the left-hand side.

Exercise 3.1. The sizes of fires. Write a program that implements a forest with $N \times N$ cells and periodic boundary conditions. Start with $N = 16$, a tree growth parameter of $p = 0.01$, and a lightning strike probability of $f = 0.2$. Make sure your code produces a graphic output of the simulation at each time step, similar to what you see in figure 3.2(a).

- **a.** When a lightning bolt strikes a cell occupied by a tree, determine the cluster to which the tree belongs and set the cells of the burned cluster to empty.
- **b.** Visualize the time evolution of the forest. Play with the parameters p and f to determine their effect. Find parameters that create very small fires, and parameters that result in large fires.
- **c.** Increase N (e.g., $N = 256$) and run the simulation for $T \sim 10^4$ time steps. Each time a lightning bolt strikes a tree and starts a fire, record its size (i.e., the number of trees that are burned down in the fire). Make a histogram of the fire size distribution. Do you notice anything interesting?

3.2 Power-law behavior

In exercise 3.1 we implemented a program that, given the size N of the forest, the tree growth probability p, and the lightning strike probability f, calculates the size n_j of the fires occurring in the forest with $j = 1, \dots, k$, where k is the total number of fires. We now want to study the distribution of the fire sizes. One way to do this would be to plot the *probability distribution* $p(n)$, where n is a positive integer representing the fire size (more precisely, $n \in [1, N^2]$ for an $N \times N$ forest). However, this approach proves to be non-optimal, because very large fires are extremely rare and, therefore, we would need to perform an unreasonably long simulation. In more mathematical terms, this is due to the fact that the probability distribution of the fire size follows a power law, i.e., $p(n) \propto n^{-\alpha}$ with $\alpha > 1$ for $n \to \infty$. Furthermore, the finite size of our arena combined with the periodic boundary conditions might also be a problem, because it induces finite-size effects in the fire size distribution (in particular, for fires that are close to their maximum size N^2).

Power-law distributions have noisy tails and are data hungry. Instead of directly computing the probability distribution of fire sizes $p(n)$ (which might be very noisy if k is small), we are going to calculate the *complementary cumulative distribution function* (cCDF) $C(n)$, i.e., the probability of a fire size being larger than or equal to n. More explicitly, $C(n)$ can be expressed as follows:

$$C(n) = \sum_{m=n}^{+\infty} p(m). \tag{3.1}$$

As $p(n) > 0$ for all n (i.e., fires of all size are possible), $C(n)$ is positively defined and monotonically decreasing, taking values from one to zero. If the probability function has a power-law distribution (i.e., $p(n) \propto n^{-\alpha}$), then

$$C(n) \propto n^{1-\alpha} \qquad \text{for} \qquad n \to \infty. \tag{3.2}$$

Exercise 3.2. Power-law probability distributions. Given a probability function $p(n) \propto n^{-\alpha}$ for $n \geqslant n_0$, show that the cCDF $C(n) \propto n^{1-\alpha}$ for $n \geqslant n_0$. *[Hint: Use the definition of* $C(n)$ *given in equation (3.2).]*

Using $C(n)$, we can easily (and efficiently) generate a series of random numbers n_i that are distributed according to $p(n)$. To do so, we need to take $C^{-1}(n)$, i.e., the inverse of $C(n)$, which is always well-defined because $C(n)$ decreases monotonically from one to zero. By generating a uniformly distributed sequence of numbers r_i in the interval [0, 1], we can directly build the sequence n_i by setting $n_i = C^{-1}(r_i)$, which is distributed according to the probability function $p(n)$ (when implementing this, you need to consider that $C(n)$ is a discrete function and therefore r_i should be a value of its image, which can easily be achieved by rounding).

Exercise 3.3. Random numbers with a power-law distribution. Implement a function that, given the exponent $\alpha > 1$, generates a sequence of positive random integers $n_i > n_{\min}$ distributed according to the probability $p(n) \propto x^{-\alpha}$. *[Hint: Follow the procedure described in the paragraph above: write* $p(n)$*; calculate* $C(n)$*; invert it to find the analytical dependence for* $C^{-1}(n)$*, and use it to generate the sequence. You might execute this exercise with continuous numbers instead of integers, in which case you will obtain* $x_i = x_{\min} r_i^{\frac{1}{1-\alpha}}$*.]*

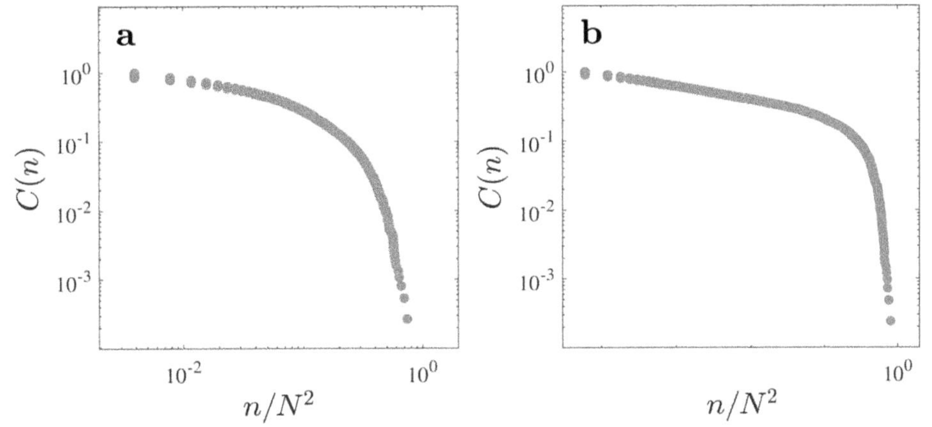

Figure 3.3. Power-law distribution for the forest fires. (a) cCDF $C(n)$ (equation (3.1)) as a function of the relative fire size (n/N^2) for a small forest ($N = 16$) and (b) for a larger forest ($N = 256$). Note that the linear section of the plot becomes longer as N increases. The parameters used for these simulations are $p = 0.01$ and $f = 0.2$.

In real-life situations, we need to determine the probability function $p(n)$ from a limited series of observations. For example, when analyzing forest fires, we have a sequence of empirical fire size values n_j with $j = 1, \ldots, k$, where k is the total number of fires from which to estimate $C(n)$ and $p(n)$ (see also exercise 3.1). To determine $C(n)$, we can do the following:

1. Sort the fire sizes array $\{n_j\}$ into ascending order, building the new sequence s with $s_i \leqslant s_{i+1}$ for all i.
2. Create an array $C(s)$ that decreases monotonically: $\left[\frac{k}{k}, \frac{k-1}{k}, \ldots, \frac{3}{k}, \frac{2}{k}, \frac{1}{k}\right]$. As all elements in s after s_i are greater than or equal to s_i, C_i gives the probability that an element in s is larger than or equal to s_i.
3. To visualize the cCDF as a function of the fire size, plot $C(s)$ as a function of s.

On a logarithmic scale, we obtain linear behavior as a result of the power-law distribution, as presented in figure 3.3. This linearity is distorted because of finite-size effects (i.e., in a finite forest of $N \times N$, the fire sizes have an upper bound of N^2).

Exercise 3.4. Power-law distribution of the forest fires. Simulate forest fires with ($p = 0.01$, $f = 0.2$). Record the sizes of at least 5000 fire events in order to study their distribution.

a. Calculate and plot the complementary cCDF $C(n)$ for $N = 16$ versus the relative fire size (n/N^2), as shown in figure 3.3(a). Explain the deviation from linear behavior.

b. Repeat the same simulation for $N = 256$, as shown in figure 3.3(b). Show that the linear section of the plot becomes longer as N increases.

Exercise 3.5. Fire-grown forests vs random forests. We are now ready to use the concept of the empirical cCDF to exemplify the difference between fire events in a forest that has grown with fires and a randomly generated forest. Starting from the code

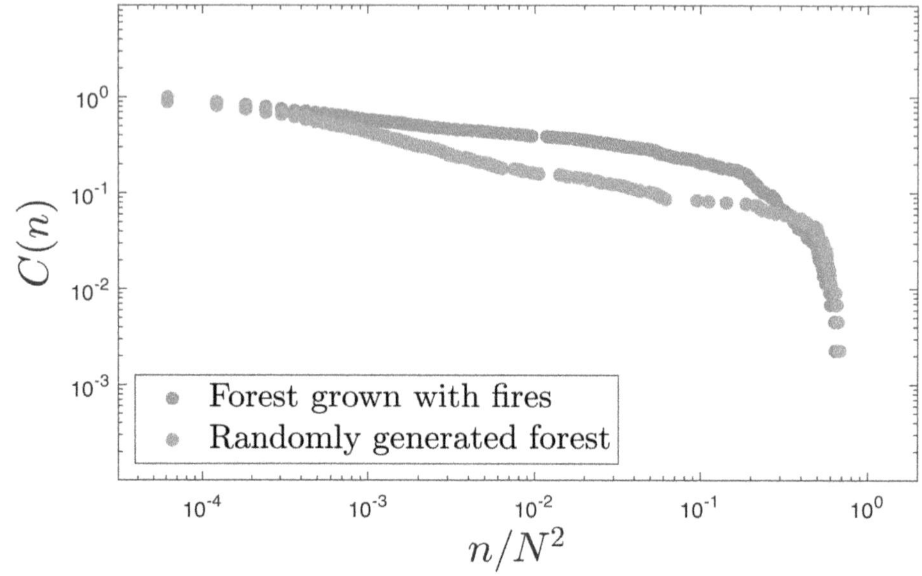

Figure 3.4. Fire-grown vs random forests. Fire size distribution in a fire-grown forest (blue dots), compared with a randomly grown forest (orange dots). When a fire occurs, the size of the forest right before the fire is recorded. A random forest with the same number of trees is generated, a fire is ignited at a random place occupied by a tree, and the size of the fire in the random forest is recorded. The figures have been obtained using a numerical simulation with $p = 0.01$, $f = 0.2$, $N = 128$, and 1000 fires in each case. For this choice of parameters, the difference between the two distributions is evident.

implemented in exercise 3.1, choose some values for N (e.g., $N = 128$), p, f, T. Start the evolution of your forest and record the fire sizes. Each time a fire occurs, generate a random forest that has the same number of trees as the original forest before the fire, start a fire in this randomly generated forest and record its size. After completing T time steps, we have had k fires for each case.

a. Compare the cCDFs of the fire sizes obtained for the original forest that is grown with repeated fires and the randomly generated forest. Compare your results with figure 3.4.

b. Play with the parameters p, f, T. Repeat the exercise, and check the effects of your choices of p, f, T on the cCDF plot. What do you observe?

We have seen in exercise 3.5 that the distribution of fire sizes in a forest grown with recurring fires differs greatly from that in a forest with the same density, but grown randomly without fires. The self-organized fire distribution follows a power law with an exponent of $\alpha = 1.15$ [4] (i.e., $p(n) \propto n^{-1.15}$) in the limit of large forests. In the following exercise, we are going to evaluate this claim. We will not follow the method of a maximum likelihood estimator (MLE) to estimate the exponent [5] because, in this case, the cutoff of our finite distribution would introduce too much error. Instead, we will follow a less refined though still effective approach to estimate

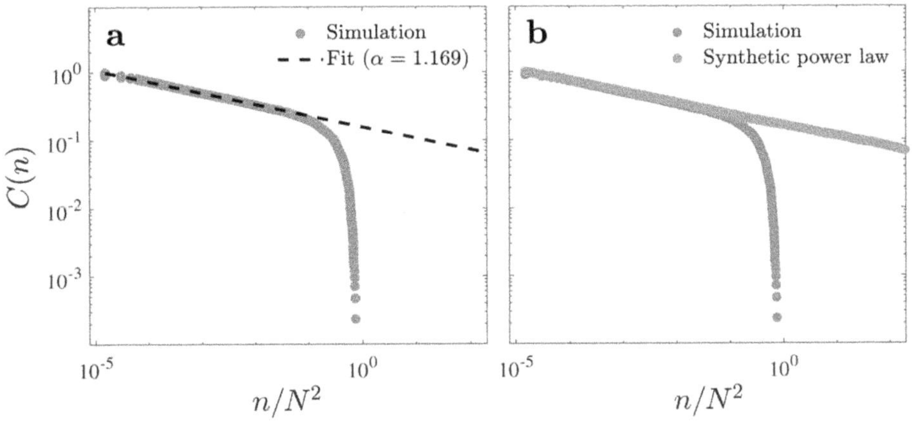

Figure 3.5. Fitting the fire size distribution to a power law and a comparison with power-law-distributed synthetic data. (a) cCDF $C(n)$ as a function of the relative fire size of a forest fire simulation with parameters $p = 0.01$, $f = 0.2$, and $N = 256$. The linear trend, visible for the smaller fire sizes, gives a slope of $\beta = -0.169$ in the log–log plot. This corresponds to a power-law probability distribution of $P(n) \propto n^{-\alpha}$ with $\alpha = 1 - \beta = 1.169$. (b) The data obtained from a forest fire simulation (blue dots) are compared with a set of synthetic fire sizes (red dots) with a power-law distribution using the exponent α obtained from the linear fit. The two data sets have the same initial trend in the plot but deviate from each other for larger fire sizes, as the maximum relative fire size in a simulation (n/N^2) is 1. As expected, the slope of the power-law-distributed synthetic data remains the same over the full range.

the exponent of the probability distribution. If $p(n) \propto n^{-\alpha}$ for $n \geqslant n_0$, then $C(n) \propto n^{1-\alpha}$ for $n \geqslant n_0$, and, therefore,

$$\log(C(n)) = (1 - \alpha)\log(n) + c \tag{3.3}$$

for $n \geqslant n_0$. Plotting $\log(C(n))$ vs $\log(n)$, the trend is linear with a slope of $\beta = (1 - \alpha)$. Fitting the slope β allows us to obtain $\alpha = 1 - \beta$. However, we do not have the theoretical $C(n)$, but the empirical $C(j)$ expressed as a function of the relative fire size. A linear fit of the slope of $\log(C(n))$ as a function of $\log(n/N^2)$ for the intermediate values of the domain $[1/N^2, 1]$ would give us the slope β and thus the coefficient α. Note that the linear fit is better when made in the linear section of the plot (as demonstrated in figure 3.5(a)), as the results deviate from the theory for larger fire sizes due to the upper bound of the finite forest size.

Exercise 3.6. Determination of the exponent α. Starting from your code for exercise 3.5, choose some values for N (e.g., $N = 256$), p, f, T.

a. Generate the sequence of growth and fires, and record the fire sizes. Plot $C(n)$ vs the relative fire size n/N^2.

b. Fit the slope of $\log(C(n))$ versus $\log(n/N^2)$ in the initial part of the distribution (neglect the part close to 1 because of finite-size effects). From the slope β, calculate $\alpha = 1 - \beta$, the exponent of the power-law distribution. Compare the cCDF plot and the linear fit with figure 3.5(a).

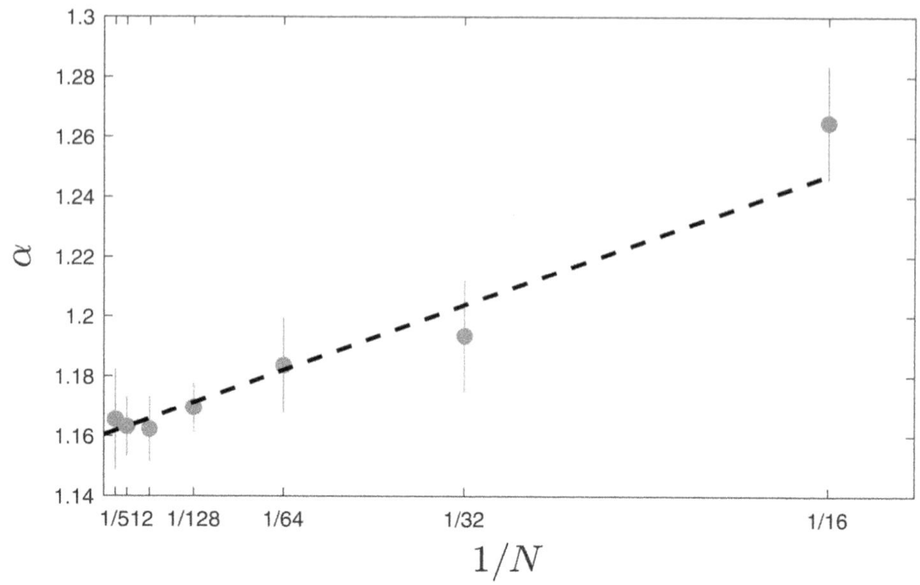

Figure 3.6. Dependence of the power-law distribution exponent as a function of the inverse of the forest size. The graph shows a possible way to extrapolate the value of the exponent α to the case of $N \to \infty$. We perform numerical simulations for a certain set of N values. For this figure, $N = 16, 32, 64, 128, 256, 512, 1024$ have been used, and for each value of N ten different numerical simulations have been performed and their characteristic exponent α recorded, to obtain an estimate of its average $\langle \alpha \rangle$ and its standard deviation $\Delta\alpha$. The linear fit is made using the leftmost five data points and this gives us $\alpha_\infty = 1.160 \pm 0.004$, which is not too far from the expected value of 1.15.

c. Compare the value you obtain with the reference value from the literature, $\alpha = 1.15$ [4].

d. Change the parameters p and f and explain how they affect the exponent.

Exercise 3.7. Comparison with synthetic data with a power-law distribution. Starting from your code of exercise 3.4, choose some values for N (e.g., $N = 256$), p, f, T.

a. Generate the sequence of growth and fires, and record the fire sizes. Produce a $C(n)$ vs n/N^2 plot. Make a fit to determine α, as you did in the previous exercise.

b. Generate a synthetic set of data with a distribution of $p(n) \propto n^{-\alpha}$ according to exercise 3.3. Produce the $C(n)$ vs n/N^2 plot for this case.

c. Compare the results from the previous two points, as shown in figure 3.5(b). What do you notice for the relative fire sizes that are close to one?

d. Change the parameters p and f and explain how they affect the comparison with the synthetic power-law distribution.

Exercise 3.8. Exponent α for large N. At this point, we are ready to try to estimate the value of the exponent α in the case of very large N, even beyond what we can simulate with our computational resources. Starting from your code of exercise 3.6, choose values for p, f, T.

a. Set N equal to 16, 32, 64, 128, 256, 512, 1024. For each case, run a simulation and find the value of the exponent α. In order to reduce noise, repeat each case ten times and take the average α. Let us call these different numbers α_{16}, α_{32}, and so on.

b. Plot the resulting exponents α_N as a function of N^{-1}, as shown in figure 3.6. Extrapolate your results to $N^{-1} \to 0$ (i.e., $N \to \infty$). Explain the reason for your choice of the fitting function and provide your extrapolation for α_∞. Compare your plot and your fitting procedure with figure 3.6.

c. How does the value estimated in the previous point relate to the estimate of $\alpha = 1.15$ given in reference [4]?

3.3 Further reading

This chapter is about a simple model for forest fires. Reference [6] provides more information about this model and its consistency with experimental data. Interestingly, state-of-the-art approaches to modeling and predictions of forest fires are based on the machine learning method called (rather confusingly) the *random forest* [7]. This technique, used in fields from economics to geosciences, permits one to perform classification and regression tasks and generates reasonable predictions across a wide range of data, such as those originated from experimental observations of a natural phenomenon over several years and locations in space. (In the random forest algorithm, a *tree* is a *decision tree*, and not the organic tree we meant in this chapter.) An example of a forest fire study based on random trees is given in reference [8].

For a review of power-law distributions (e.g., such phenomena in biology, demography, computer science, economics, information theory, language, and astronomy), we refer the reader to [9]. Some references that cover power-law statistics in natural disasters are [6] (forest fires) and [10] (earthquakes). Power-law frequency-size distributions are often associated with self-organized critical behavior, which can be found in a number of different models, e.g., the sandpile model [11], the slider-block model [12, 13], and the forest fire model [14].

In exercise 3.8, the basic assumption used in the estimate is that the behavior for large N can be inferred and generalized from the behavior at smaller (but large enough) N. This is what actually happens in the majority of cases: increasing the system size merely leads to more significant digits in the critical parameters. This model is different though. Reference [15] provides extensive simulations for values of N up to $N = 65536$ (by representing each tree using a single bit, you can fit such a lattice into 1 GB of memory); the authors found new behavior for extremely large systems (e.g., $\alpha = 1.19 \pm 0.01$), noting that nothing prevents further new behaviors at even larger scales.

3.4 Problems

Problem 3.1. Forest fires without a periodic boundary condition. Repeat the exercises in this chapter without periodic boundary conditions. Observe whether there are significant differences in the results of the simulations.

Problem 3.2. Forest fires in a rectangular forest. Repeat the exercises in this chapter using a long rectangle instead of a square, with and without periodic boundary conditions. Do the results change?

Problem 3.3. Forest fires in a hexagonal lattice. What happens if, instead of a square lattice, we place our trees in a hexagonal lattice?

Problem 3.4. Forest fires in 3D. Repeat the exercises in this chapter in three dimensions with periodic boundary conditions. This might be the model for a fire in a building, propagating from room to room across and between floors.

Problem 3.5. Forest fires in a random graph. Repeat the exercises in this chapter using a random graph, of the kind that resemble a geographical map of municipalities in a region. For example, the graph can be generated by connecting N random points according to their Delaunay triangulation.

Problem 3.6. Moore neighborhood. What happens in a forest set on a square lattice if, instead of the von Neumann neighborhood, we take the neighborhood consisting of the eight neighboring sites (the *Moore neighborhood*)?

Problem 3.7. Slow fire. In this chapter, we have supposed that the fire propagates immediately and burns down the cluster of trees in a single time step. Assume now that the fire propagates only to the first neighbors in each time step (the size of the fire is determined when the original burning cluster is extinguished). Study this case using the methods seen in this chapter.

Problem 3.8. Wind effect on fire propagation. Assuming the fire does not propagate instantaneously (problem 3.7), simulate the effect of wind. In fact, a steady strong wind can affect the direction of propagation of a fire; a hot dry wind can speed up the burning process and affect a larger area.

Problem 3.9. Obstacles. Assume that there are areas that cannot burn (e.g., areas with sand, rocks with no vegetation, or lakes). How does this affect the propagation of a fire? Are defined geometries helpful to prevent the spread of a fire, without limiting the amount of green areas too much?

Problem 3.10. Differently flammable vegetation. Assume that not all trees have the same probability of catching fire. For example, older trees might be more prone to fires when hit by lightning. How does this affect the results of the model?

3.5 Challenges

Challenge 3.1. Simulations for fire prevention and containment strategies. Use forest fire simulations to develop fire prevention and containment strategies and to evaluate their performance.

Challenge 3.2. Simulation of the Yellowstone fire. Simulate some forest fires that actually happened, taking into account topography, wind conditions, etc. For example, consider the (in)famous Yellowstone fire.

Challenge 3.3. Sandpile model. Implement the sandpile model (see reference [11]) with tools similar to those used in this chapter.

Challenge 3.4. Fires on fractals. Simulate and study a forest fire occurring on a fractal space. How do its properties differ from those studied in a 2D space in this chapter?

Challenge 3.5. Beyond forest fires. Can you apply the ideas gathered while studying forest fires (e.g., power laws, cell-based models) to other fields, such as economics and ecosystems?

References

[1] Bak P, Tang C and Wiesenfeld K 1987 Self-organized criticality: an explanation of 1/*f* noise *Phys. Rev. Lett.* **59** 381

[2] Jeldtoft Jensen H 1998 *Self-Organized Criticality—Emergent Complex Behavior in Physical and Biological Systems* (Cambridge: Cambridge University Press)

[3] Drossel B and Schwabl F 1992 Self-organized critical forest-fire model *Phys. Rev. Lett.* **69** 1629

[4] Grassberger P 1993 On a self-organized critical forest-fire model *J. Phys. A Math. Theor.* **26** 2081

[5] Clauset A, Shalizi C R and Newman M E J 2009 Power-law distributions in empirical data *SIAM Rev.* **51** 661

[6] Malamud B D, Morein G and Turcotte D L 1998 Forest fires: an example of self-organized critical behavior *Science* **281** 1840

[7] Breiman L 2001 Random forests *Mach. Learn.* **45** 5

[8] Tonini M *et al* 2020 A machine learning-based approach for wildfire susceptibility mapping. The case study of the Liguria region in Italy *Geosciences* **10** 105

[9] Pinto C M A, Lopes A M and Machado J A T 2012 A review of power laws in real life phenomena *Commun. Nonlinear Sci. Numer. Simul.* **17** 3558

[10] Geller R J 1997 Earthquake prediction: a critical review *Geophys. J. Int.* **131** 425

[11] Bak P, Tang C and Wiesenfeld K 1988 Self-organized criticality *Phys. Rev.* A **38** 364

[12] Burridge R and Knopoff L 1967 Model and theoretical seismicity *Bull. Seismol. Soc. Am.* **57** 341

[13] Carlson J M and Langer J S 1989 Mechanical model of an earthquake fault *Phys. Rev.* A **40** 6470

[14] Bak P, Chen K and Tang C 1990 A forest-fire model and some thoughts on turbulence *Phys. Rev.* A **147** 297

[15] Grassberger P 2002 Critical behaviour of the Drossel–Schwabl forest fire model *New J. Phys.* **4** 17

Chapter 4

The Game of Life

How can we define life? There is no simple or unanimous definition of life. All living systems share some characteristics, though: they exhibit complex emergent behaviors, can replicate themselves, and possess self-regulating mechanisms that avoid exponential growth or extinction.

In the 1940s, John von Neumann defined life as an entity that can reproduce itself and simulate a Turing machine. Together with Stanislaw Ulam, he developed the concept of the cellular automaton, which is a grid of cells, each with a definite state (e.g., on or off, one or zero) that changes at every time step on the basis of simple, fixed rules that are equal for all cells and applied simultaneously to the whole grid [1]. Later research on cellular automata by Stephen Wolfram classified them into four different categories [2]: cellular automata that evolve toward homogeneity (class 1), those that

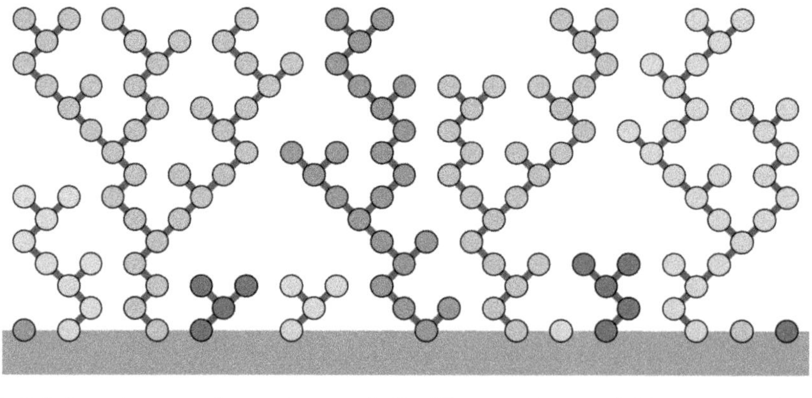

Figure 4.1. Cellular automata and emerging complexity. The stunted trees of this image are, in reality, the evolution of the so-called Rule 90, a one-dimensional cellular automaton. Though simple, even one-dimensional cellular automata can give rise to complex emergent behavior, and can be interpreted as representing the evolution of several natural systems, like the growth of a forest of stunted trees. *Source: Rule_90_trees, Wikipedia: David Eppstein.*

evolve toward an oscillating or periodic condition (class 2), those that display chaotic behaviors (class 3), and those that display complex emergent behaviors (class 4).

In 1970, John H. Conway was experimenting with two-dimensional cellular automata and testing different sets of rules. He was looking for a set of rules that could result in unpredictable complex behavior. The rules he was seeking were to have the following characteristics: they should not give rise to exponential growth, they should allow configurations with unpredictable chaotic outcomes, they should potentially satisfy the the von Neumann conditions for a universal constructor (i.e., to be a self-replicating machine), and they should be the simplest possible rules. The game was first published in 1970 in 'Scientific American,' in Martin Gardner's mathematical games section [3]. Since its appearance, the *Game of Life*, as it came to be known, has attracted a lot of interest and has been thoroughly studied. It has shown features of the emergence of complex behavior and self-organization. It was later shown to be Turing complete and capable of simulating a universal constructor or any other Turing machine. Researchers and enthusiasts are actively progressing this research today. It was even recently featured in the 'New York Times' [4] on the occasion of the fiftieth anniversary of the Game of Life and six months after the death of John Conway, who was about 80 years old and who died of COVID-19.

In this chapter, we first introduce a simple one-dimensional cellular automaton, which has the advantage of having sets of rules that can be enumerated from 0 to 255; we then focus our attention on some interesting rules (namely, Rule 184, Rule 30, Rule 90, and Rule 110, figure 4.1), mentioning their properties and applications. We then introduce the definition and properties of the Game of Life, studying its evolution in a few paradigmatic classical cases. We investigate some variations in the rules to understand how sensitive the model is and how different are the outcomes it produces. We also introduce and simulate another two-dimensional cellular automaton, the *majority rule*, which is used in the study of elections where voters must choose between two candidates, or in the modeling of markets with two competing products.

Example codes: Example Python scripts related to this chapter can be found at: https://github.com/softmatterlab/SOCS/tree/main/Chapter_04_Game_of_Life. Readers are welcome to participate in the discussions related to this chapter at: https://github.com/softmatterlab/SOCS/discussions/12.

4.1 One-dimensional cellular automata

One-dimensional cellular automata have been systematically studied by Stephen Wolfram [2]. Even though limited to a single dimension and apparently simple, they feature remarkable properties and have found important applications [5–7]. The universe of a one-dimensional cellular automaton is composed of cells arranged on an infinite line. Each cell has two neighbors, on the left and the right. There are, therefore, only $2^3 = 8$ possibilities when considering the state of a cell and its two nearest neighbors. Starting from a given initial configuration, i.e., the *parent generation*, the states of the cells evolve generation after generation. Each automaton is defined by the rules used to determine the state of the central cell, given the state of the cell itself and those of its first neighbors; therefore, there are only $2^8 = 256$ different one-dimensional cellular automata. Each of these one-dimensional cellular automata can be identified by a decimal number from 0 to 255 corresponding to the final outcomes of the action of the automaton, when acting, respectively, on the basis of states 111, 110, 101, 100, 011, 010, 001, 000. Often, the sequence of generations for a one-dimensional cellular automaton is represented using a two-dimensional diagram. Each row represents a generation, starting from the parent generation set in the top row and proceeding downwards with generations 1, 2, 3, and so on (see figure 4.2 for some example representations). In this section, we will describe four of the most interesting one-dimensional cellular automata, those known as Rule 184, Rule 90, Rule 30, and Rule 110.

Rule 184. This is an example of a class-2 cellular automaton, i.e., a cellular automaton that evolves from any initial pattern toward regular patterns that shift by one step for each generation. The automaton is defined by the following rules [8]:

Current pattern	111	110	101	100	011	010	001	000
New value for center cell	1	0	1	1	1	0	0	0

(4.1)

A typical sequence of generations for Rule 184 is given in figure 4.2(a). This set of rules can be intuitively described as follows: at each step, if a cell with a value of one has a cell with a value of zero immediately to its right, the one moves rightwards

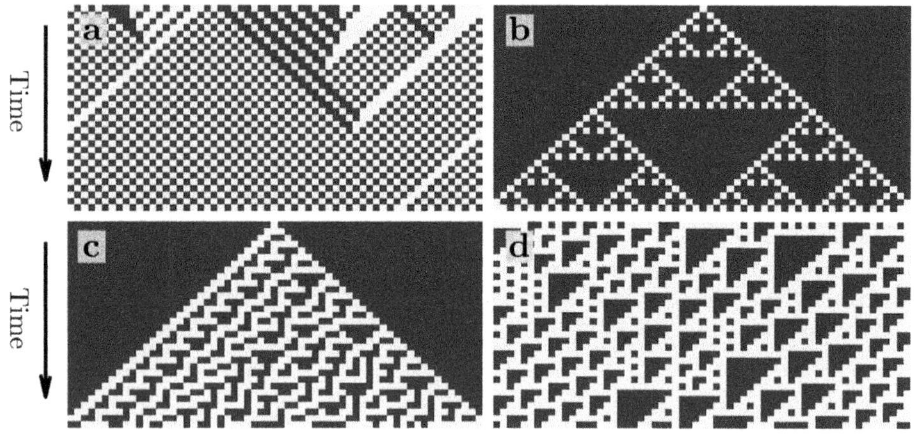

Figure 4.2. One-dimensional cellular automata. Examples of the evolution over time of cellular automata. The yellow cells represent ones, and the beige cells represent zeros. Each row represents the state of the cellular automaton at a given time, and the evolution over time can be seen by scanning the rows downwards. (a) Rule 184 is a class-2 cellular automaton, because its evolution tends to a state that moves each generation by one step as time passes (e.g., traffic flow in a single lane, particle deposition on a surface, ballistic annihilation of particles). (b) Rule 90 is a class-4 cellular automaton, because it can show self-similarity and emergent complex behavior. (c) Rule 30 is a class-3 cellular automaton, because it shows chaotic behavior. It can be used in random number generators. (d) Rule 110 is a very peculiar and unique cellular automaton. It can show emergent complex behavior as well as chaotic behavior. This cellular automaton has been shown to be equivalent to a Turing machine.

leaving a zero behind. A one with another one to its right remains in place, while a zero that does not have a one on its left stays a zero. Rule 184 can represent the evolution of several systems. For example, it can provide a simple model for the traffic flow in a single lane (Figure 4.3(a)), for the deposition of particles on an irregular surface (Figure 4.3(b)), and for the ballistic annihilation of particles in one dimension (figure 4.3(c)).

Rule 90. This is an example of a class-4 cellular automaton, which gives rise to emergent complex behavior. It is defined by the following rules [9]:

Current pattern	111	110	101	100	011	010	001	000	(4.2)
New value for center cell	0	1	0	1	1	0	1	0	

The state of each cell is determined solely by the states of its neighbors, disregarding the state of the cell itself. This rule actually implements an XOR (exclusive or) of the

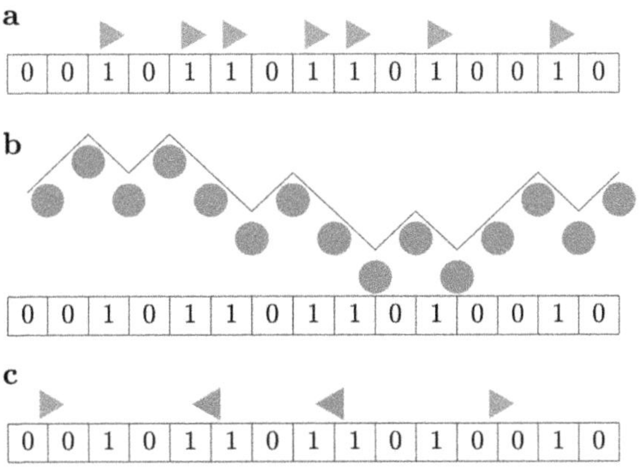

Figure 4.3. Rule 184. Rule 184 can be used to represent the time evolution of systems with simple one-dimensional dynamics, such as (a) traffic flow in a single lane; each onstance of one represents a car trying to go to the right. Only cars with an empty space in front of them are able to move. (b) Particle deposition: the zeros and ones represent the boundary of a surface on which particles (blue circles) are deposited. The boundary follows the profile of particle accumulation, which rises and falls. In the next generation, new particles can only be deposited on the local minima of the boundary, and the boundary is updated accordingly. (c) Ballistic annihilation: contiguous regions with multiple zeros represent a particle going to the right, while contiguous regions with multiple ones represent a particle going to the left. At each time step, the particles move by one cell. When they meet, they annihilate, generating an empty space (a sequence of alternating single ones and zeros). Interestingly, the ballistic annihilation interpretation can also be used for content-free language parsing: an open parenthesis is a particle traveling to the right, and a closed parenthesis is a particle traveling to the left. A well-structured expression contains an equal number of open and closed parentheses; therefore, it is possible to check the syntax of the parentheses in a regular expression by letting its representation evolve with rule 184, and see if it arrives at a state in which all parentheses are annihilated.

state of the neighboring cells. It can describe the growth of a forest of stunted trees (see figure 4.1). It also features self-replication: for example, when starting from an initial condition with a single one, the time evolution of Rule 90 leads to a fractal pattern (the *Sierpinski triangle*, see figure 4.2(b)).

Rule 30. This is an example of a class-3 cellular automaton, whose evolution leads to a chaotic pattern. This cellular automaton is defined by the following rules [10]:

Currentpattern	111	110	101	100	011	010	001	000
New value for center cell	0	0	0	1	1	1	1	0

(4.3)

A typical sequence of generations for Rule 30 is shown in figure 4.2(c). This cellular automaton exhibits features of aperiodicity and chaotic behavior, which have led to its application in random number generation and cryptography [11]. Furthermore, its time evolution represented in two dimensions resembles the patterns found on shells, such as those of *Conus textile* sea snails.

Rule 110. This unique cellular automaton can be considered a class-4 cellular automaton, but it also shares features of a class-3 cellular automaton. It is defined by the following rules [12]:

Current pattern	111	110	101	100	011	010	001	000
New value for center cell	0	1	1	0	1	1	1	0

(4.4)

This cellular automaton is unique because of its behavior at the boundary between stability and chaos, and for its characteristic of *universality*, i.e., it is Turing complete, which means that it can be used to simulate any Turing machine, as demonstrated by Matthew Cook [13].

Exercise 4.1. One-dimensional cellular automata. Write a program that implements the one-dimensional cellular automata described in the text above. Given the number of the rule, it should produce an animation of the corresponding cellular automaton.

- **a.** Run the simulation for the cellular automata described in the text, namely Rule 184, Rule 90, Rule 30, and Rule 110. *[Hint: Compare your outcomes with figure 4.2.]*
- **b.** Experiment with other rules! Try to find cellular automata corresponding to each of the four classes identified by Stephen Wolfram.

4.2 Conway's Game of Life

The set of rules of Conway's Game of Life are pretty simple [3]. On an infinite two-dimensional board, the cells have only two states, one (or *on*, *alive*) and zero (or *off*, *dead*). Initially, at generation zero, a certain number of cells are alive and the rest are dead. The next generation is calculated from the current generation by applying the following rules:

	Number of live neighbors	Next generation
Dead cell	0, 1, 2	remains dead
	3	a new live cell is born
	4, 5, 6, 7, 8	remains dead

(4.5)

	Number of live neighbors	Next generation
Live cell	0, 1	dies (loneliness)
	2, 3	remains alive
	4, 5, 6, 7, 8	dies (overpopulation)

When considering the number of neighbors, the *Moore neighborhood* is employed (figure 4.4(a) and 4.4(b)).

Exercise 4.2. Game of life. Implement the Game of Life on a square 10×10 grid. Start from a random configuration and proceed for 20 generations.

a. Structure your program using a function that, for every generation, finds the number of live neighbors for each cell, and a function that calculates the next generation on the basis of the rules (4.5). Assume non-periodic boundary conditions (i.e., a cell at the boundary has less than eight neighboring cells, as shown in figure 4.4(a)).

b. Visualize your output for each generation (figure 4.4(c)–(d)) and check that every cell is updated correctly from one generation to the following one.

c. Modify your program to find the number of neighbors of each cell for the case of periodic boundary conditions (figure 4.4(b)).

Let us start by applying the rules (4.5) to some simple configurations. If, for example, we start from the *block*, shown in figure 4.5(a), we find that the next generation is exactly equal to the parent generation. Shapes that have this property are called *still lifes*. In figure 4.5 you can find the most common still-life examples.

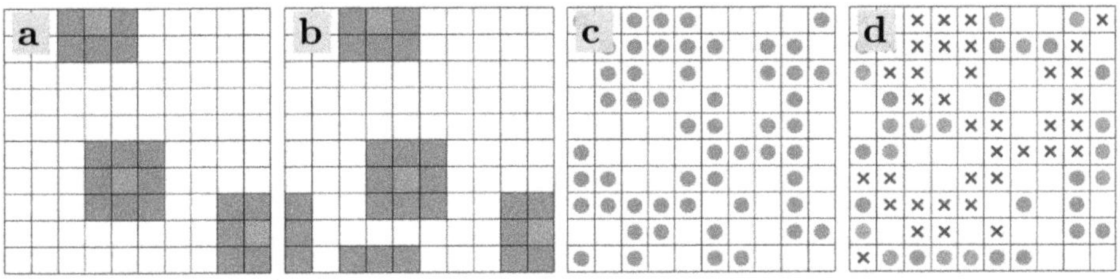

Figure 4.4. Game of life rules. A cell (orange) and its first Moore's neighbors (cyan) (a) without and (b) with periodic boundary conditions. (c) Example of a generation and (d) of the subsequent generation obtained by applying the rules (4.5): cells with crosses have died, cells in orange have been newly born, and cells in cyan survived from the previous generation.

Exercise 4.3. Time evolution of a still life. For each still life represented in figure 4.5, write a function that reproduces its time evolution.

If, instead, we start with the *blinker* (figure 4.6(a)), the next generation (figure 4.6(b)) is different from the parent, but the second generation returns to the original form (figure 4.6(c)). This group of periodical configurations is called *oscillators*.

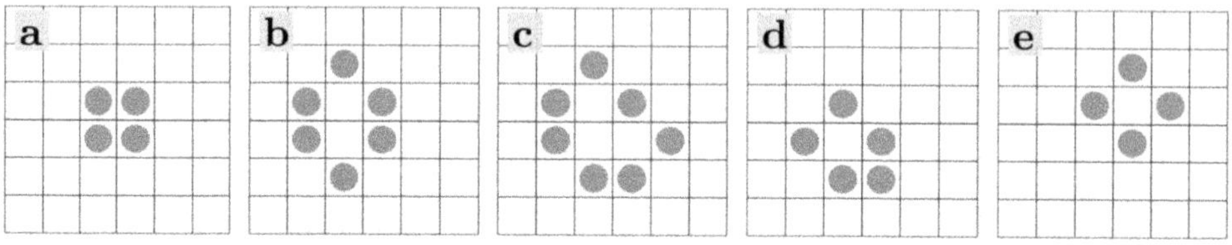

Figure 4.5. Still life. Examples of still life: (a) the block, (b) the beehive, (c) the loaf, (d) the boat, and (e) the tub. The next generation originated by these shapes does not change. Can you think of other still-life examples?

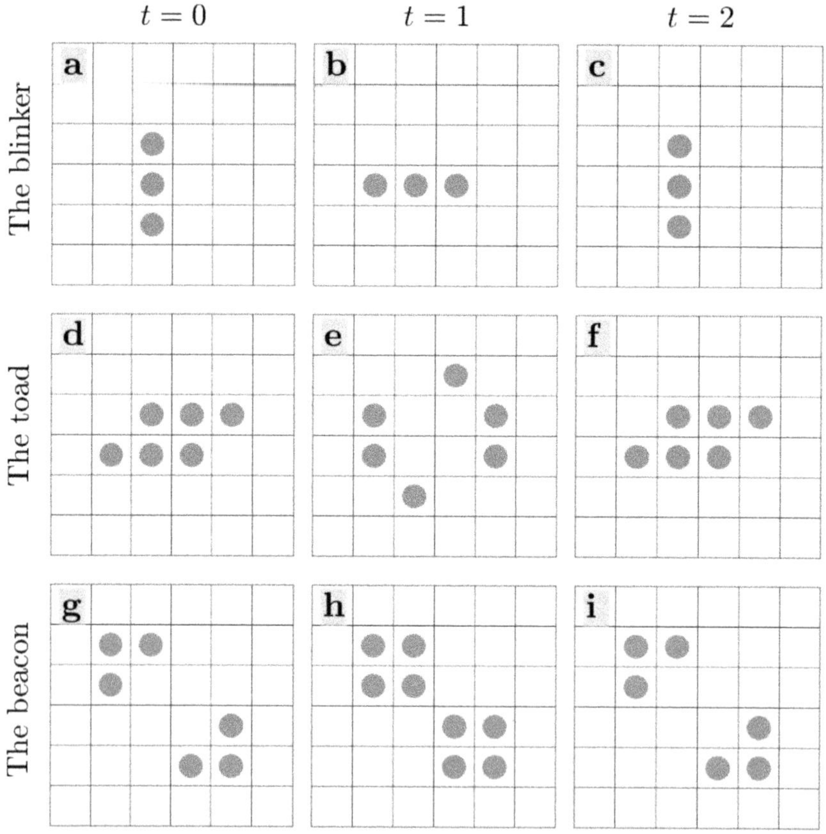

Figure 4.6. Oscillators. Examples of oscillators: (a)–(c) the blinker, (d)–(f) the toad, and (g)–(i) the beacon. Three generations are shown, indicated by the timestamp. After two generations, the initial shape repeats itself.

Figure 4.6 represents the most common oscillators with a period of two. There are also oscillators with periods greater than two. Can you think of some examples?

Exercise 4.4. Time evolution of an oscillator. For each oscillator represented in figure 4.6, write a function that reproduces its time evolution.

Still lifes and oscillators are configurations that do not move. The simplest configuration capable of motion in the Game of Life universe is the *glider* (figure 4.7). As you can see, after four generations the glider has regained its initial form, but its position is shifted diagonally by one step. Its velocity is $v = \frac{1}{4}c$, where c is the velocity of one cell per generation. In general, if after N generations a shape has regained its original form but shifted by M cells, then its velocity is $v = \frac{M}{N}c$. A velocity of c, i.e., one cell per second, cannot be attained by any object (why? and how close to one cell per second is it possible to get?).

Exercise 4.5. Time evolution of a glider. Implement the glider represented in figure 4.7.

a. Test your function by checking the time evolution of the glider over five successive generations. In which direction does your glider move?

b. Implement gliders that move in each diagonal direction.

Gliders are not the only movable objects in the Game of Life universe. There are several other objects, called *spaceships,* that move. Each spaceship has its own period and its own velocity. Gliders, among the spaceships, have a particular importance because they are very simple objects and it has been shown that, for any stable configuration, there are a finite number of long-distance gliders that can build that configuration starting from an empty board, and there is also a constructive procedure that can be used to determine the initial configuration of the gliders [14–16].

Exercise 4.6. A quest for new oscillators and gliders. How can you find new oscillators, gliders, spaceships, contained in a square $N \times N$? One way is to take a

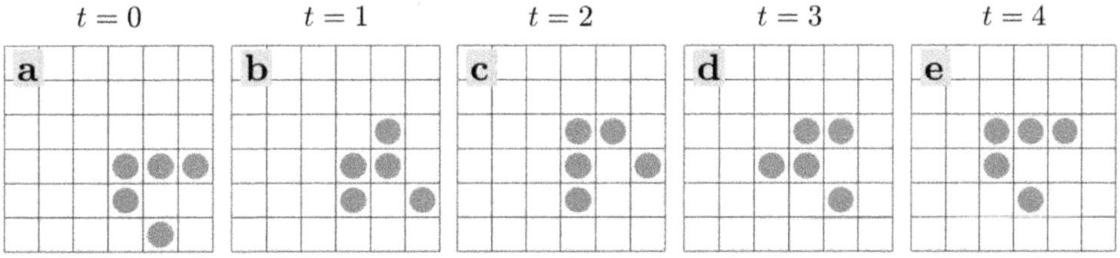

Figure 4.7. Glider. The glider regains its original shape after four generations, but it is shifted diagonally with respect to its initial position.

board with a size of $3N \times 3N$, generate a random configuration $N \times N$ at its center, and let it evolve for K generations. If generation k is the first generation that is identical to the starting configuration, then you have found an oscillator with period k. If generation k is a shifted version of the initial configuration that is shifted by s_x cells in the x direction and s_y cells in the y direction, then you have found a spaceship with velocity $v = \frac{s}{k}c$, with $s = \sqrt{(s_x^2 + s_y^2)}$.

a. Implement a function that, given a configuration, translates it by s_x cells in the x direction and s_y cells in the y direction, and a function that checks whether the two configurations are identical.

b. Implement the algorithm and use it to look for new oscillators and spaceships, starting from a random configuration.

c. How easy is it to find a new oscillator or a new spaceship using this procedure? Does a generic random configuration easily evolve into a stable one? How can you improve the algorithm to find new spaceships or oscillators?

Initially, John H. Conway was convinced that no configuration could give rise to a continuous growth of the number of live cells. However, this conjecture proved to be wrong: a team led by Bill Gosper found a configuration called the *Gosper glider gun* (figure 4.8), which produces its first glider at the fifteenth generation and then another at every 30 generations. After that discovery, several other growing configurations have been found. The taxonomy of the entities in the Game of Life includes now objects like *guns* (objects that shooting new gliders), *puffer trains* (objects that leave a trail of debris behind them), *rakes* (objects that producing spaceships), *breeders* (objects that leave behind a trail of guns), and so on. Also, the Game of Life has been shown to be Turing complete [17].

Exercise 4.7. Alternative rules for the Game of Life. Modify the rules (4.5) and explore the features of the ensuing modified Game of Life.

a. Starting from a random configuration, what does the evolution look like? Do the modified rules give rise to emergent complexity, or does the system evolve toward a stable pattern? Does the system evolve toward overpopulation or toward extinction?

b. Explore several possibilities. Find at least one non-trivial set of rules that leads to extinction and one that leads to a fixed or oscillating stable pattern.

c. How likely is that complexity emerges from a random set of rules?

4.3 Majority rule

Another interesting set of rules is the so-called *majority rule* [18]. This rule does not give rise to complex dynamics, but it has interesting applications to elections where voters must choose between two parties (or to markets where consumers must choose between two competing brands). Thus, we can suppose that each cell represents an elector, and that the states zero and one represent a vote for one of the two parties. The voting process is iterative, and each elector can reconsider their

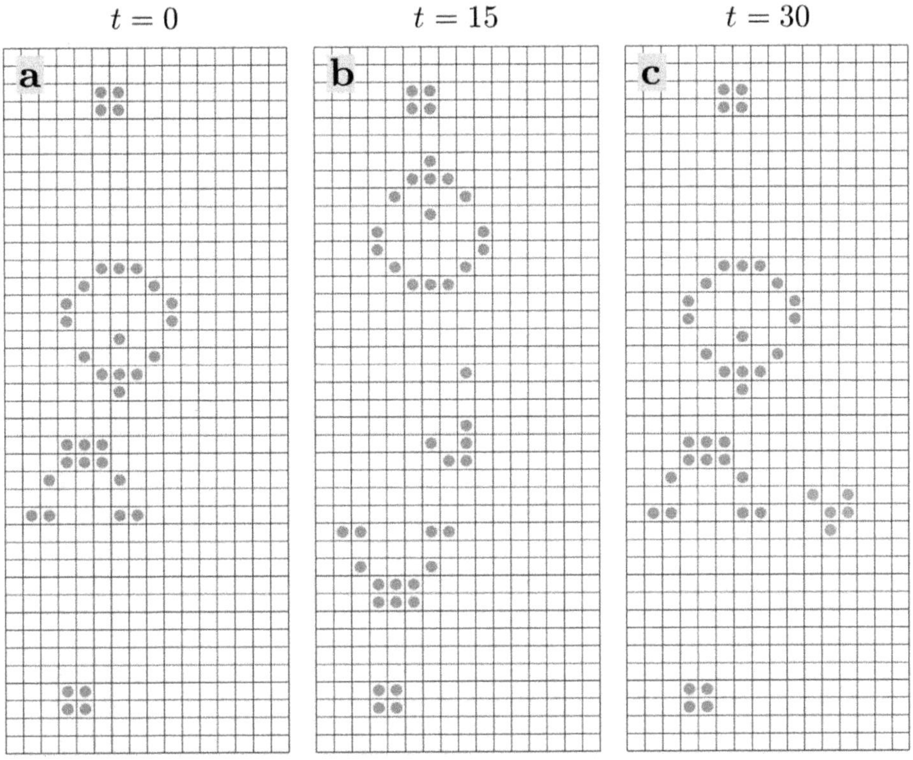

Figure 4.8. Gosper glider gun. (a) Parent generation, (b) fifteenth generation, and (c) thirtieth generation of a Gosper glider gun. At the fifteenth generation, a glider is emitted. When (a) and (c) are compared, the two configurations are identical, with the exception of the generated glider (drawn in orange in (c)), which, in the meanwhile, has been translated diagonally toward the southeast. The Gosper glider gun is ready to generate another glider at the forty-fifth generation, and the process continues forever.

vote on the basis of the votes of their neighbors. At each iteration, the electors change their vote to the vote expressed by the majority of their Moore's first neighbors. If no majority exists, then the electors preserve their former vote. The rules in this system are therefore as follows:

Number of neighbors voting 1	Next generation	
0, 1, 2, 3	vote 0	
4	vote does not change	(4.6)
5, 6, 7, 8	vote 1	

Exercise 4.8. Majority rule. Modify your code for the Game of Life by introducing the majority rule (4.6). Start with a 100×100 square board with a random configuration in which the initial fraction of 'one' votes is p. For example, start with $p = 0.45$.

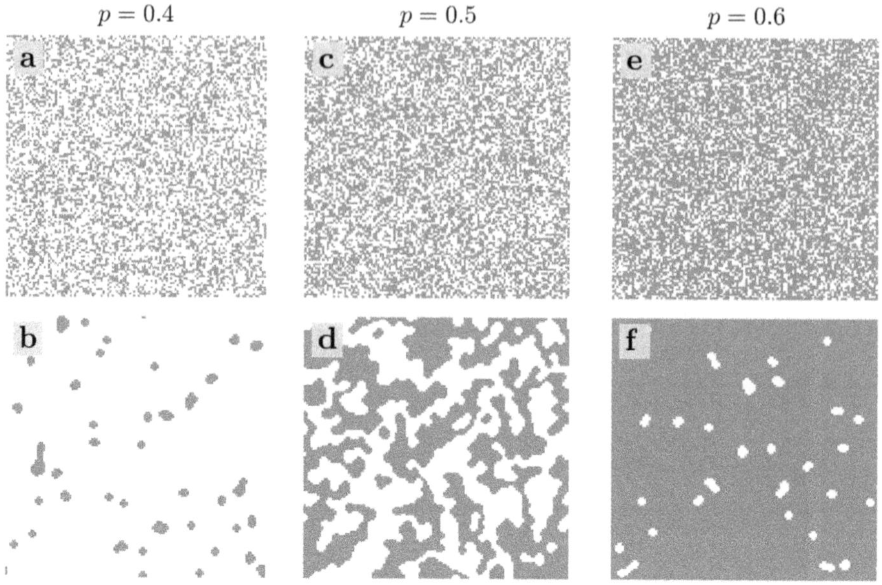

Figure 4.9. Majority rule. Initial (upper row) and final (lower row) states for a majority rule game with an initial distribution of 'one' votes with a probabilities of (a)–(b) $p = 0.4$, (c)–(d) $p = 0.5$, (e)–(f) $p = 0.6$. One can see that the vote remains balanced with $p = 0.5$. By contrast, for $p \neq 0.5$, the majority rule leads the vote to a stable outcome in which the disproportion between the two factions is much larger than the initial one.

a. Find the final state of the system: after how many iterations does the configuration become stable? What is the outcome of the vote? How are the two voter domains spatially distributed?

b. Run your simulation for different initial distributions of votes by varying p. Record the outcome of each election after the configuration stabilizes. In addition, record the number of iterations needed for stability. Compare your results with those shown in figure 4.9.

As a point related to the majority rule, it is worth mentioning the *majority problem*, which is the problem of finding a one-dimensional cellular automaton that correctly classifies a majority vote, i.e., that decides whether the majority of the votes are one or zero on the basis of local rules only. As figure 4.9 suggests, the majority rule seems to be a very good candidate for classifying the outcome of a binary election, and it works very well when the initial proportion of zeros and ones is not too close to 50%. However, it has been shown that it is not a perfect classifier [19], especially in the case of a division of votes that is very close to 50%. For example, for configurations that differ by only one vote and for which the deciding vote is surrounded by opposite votes, the deciding vote is flipped by the majority rule. Moreover, it has been shown that it is impossible to have a perfect majority classifier, either deterministic or stochastic, for any dimension [20].

4.4 Further reading

The literature on the topics of the Game of Life and cellular automata is vast. Online resources for simulations are very easily found. Among the many, it is worth mentioning the websites LifeWiki [21] and ScholarPedia [22], a website demonstrating various cellular automata [23], and a resource for the hexagonal Game of Life [24].

The first public appearance of Conway's Game of Life was in an article by Martin Gardner in Scientific American [3]. Moreover, the contributions of Stephen Wolfram are important in the fields of cellular automata, artificial life, and complexity. The highly recommended references [2, 25] provide the context of research into cellular automata, its possible applications, and its open challenges.

Regarding the historical and theoretical context of the mathematical definition of life and the foundation of the theory of self-replicating machines and cellular automata, we refer the reader to [1, 26]. It is worth reading the article about the contributions of John von Neumann [27] to the field of artificial life. This article also gives an all-round picture of the scientist and person.

4.5 Problems

Problem 4.1. Statistics of the Game of Life evolution. Study the evolution of the Game of Life.

- **a.** Start from a random configuration and let the system evolve over a large number of generations (e.g., $T = 10^4$). For each initial configuration, check whether it ends up extinct, in a still pattern, in an oscillating pattern with a period of two, or otherwise.
- **b.** Perform simulations for different starting densities of random initial patterns. Determine and analyze the statistics of the final configurations.

Problem 4.2. A modified Game of Life. The rules of the Game of Life were carefully chosen to feature complex dynamic behavior and to avoid exponential growth.

- **a.** Show that with the modified rule that particles with four neighbors also give rise to a new living particle, the game tends to evolve toward a static situation.
- **b.** Experiment with the rules: check which ones lead to extinction, saturation, or a different equilibrium condition.

Problem 4.3. A stochastic Game of Life. Modify the rules of the Game of Life so that, when a dead cell has two neighbors, there is a probability p that a new cell is born at that position.

- **a.** Study the statistical properties of the system with this stochastic set of rules, and compare them to the results of problem 4.1.
- **b.** Explore other possibilities: for example, add the possibility that a live cell with three neighbors dies. Study the statistical properties of such a system.

Problem 4.4. A hexagonal Game of Life. Instead of running the Game of Life on a chessboard-like structure, transpose its rules to simulate it on a honeycomb geometry, in which, instead of square unit cells, there are regular hexagons. *[Hint: What can seem challenging from the implementation point of view is the structure that stores information about the connections between each cell and the other cells. For a square lattice, it is straightforward to use a bidimensional array. For a hexagonal lattice, it is also possible to use a two-dimensional array, similar to the one used for a square lattice. In fact, if, instead of the hexagonal (or square) tiles, we think in terms of points, the connections to the first neighbors form a graph. The two arrangements differ in the number of first neighbors: six for the hexagonal lattice and eight for the square lattice. It is possible to represent the connections of the hexagonal lattice using the same two-dimensional array as for the square lattice. This is done by omitting the cells that are located diagonally to the upper left and the lower left from the list of the first neighbors of a cell. Then, when drawing the representation of the hexagonal lattice, we have to take care to draw the cells in neighboring rows with their centers shifted by half of the lattice distance between contiguous cells in the same row.]*

- **a.** Modify your code by writing a new procedure to determine the first neighbors of each cell, leaving out the cells that are located diagonally to the upper left and the lower left. Test your procedure. Write a function to represent your square cell array in the proper hexagonal geometry. Start from the case with no periodic boundary conditions.
- **b.** Run your simulation in this modified geometry. What are the features of the Game of Life in the new geometry? Can you find the still lifes, oscillators and gliders in this geometry? Can you characterize other relevant patterns?
- **c.** Repeat the steps above, this time considering periodic boundary conditions. Do you have a constraint on the size of the lattice?

4.6 Challenges

Challenge 4.1. Three-dimensional Game of Life. Can you find a set of rules in three dimensions that give rise to a 3D Game of Life with features similar to those of the classical Game of Life? You might first read references [28, 29].

Challenge 4.2. Game of Life with three states. Explore the possibility of defining a Game of Life in which the state of a cell can have three different values instead of two. You can be inspired by perusing reference [23].

Challenge 4.3. Game of Life with second nearest neighbors. Explore the possibility of defining a Game of Life in which the state of a cell also takes into account the information of the states of the second nearest neighbors.

Challenge 4.4. Game of Life on a graph. Simulate the Game of Life on a graph, where cells are the nodes of the graphs and their nearest neighbors are defined by the edges of the graph. Explore how the resulting properties of this modified Game of Life depend on the structure of the graph and on the set of rules. Check chapter 12 for ideas for possible graph structures to explore.

References

[1] Von Neumann J and Burks A W 1966 *Theory of Self-Reproducing Automata* (Champaign, IL: University of Illinois Press) https://archive.org/details/theoryofselfrepr00vonn_0/
[2] Wolfram S 2002 *A New Kind of Science* (Champaign, IL: Wolfram Media)
[3] Gardner M 1970 Mathematical Games—the fantastic combinations of John Conway's new solitaire game 'life' *Sci. Am.* **223** 120
[4] Roberts S 2020 The Lasting Lessons of John Conway's Game of Life (https://www.nytimes.com/2020/12/28/science/math-conway-game-of-life.html)
[5] Wolfram S 1983 Statistical mechanics of cellular automata *Rev. Mod. Phys.* **55** 601
[6] Wolfram S 1984 Cellular automata as models of complexity *Nature* **311** 419
[7] Wolfram S 1986 Cellular automaton fluids 1: basic theory *J. Stat. Phys.* **45** 471
[8] Wolfram S rule 184 - Wolfram—Alpha (https://wolframalpha.com/input/?i= rule+184)
[9] Wolfram S rule 90 - Wolfram—Alpha (https://wolframalpha.com/input/?i=rule+ 90)
[10] Wolfram S rule 30 - Wolfram—Alpha (https://wolframalpha.com/input/?i=rule+ 30)
[11] Wolfram S 1986 Random sequence generation by cellular automata *Adv. Appl. Math.* **7** 123
[12] Wolfram S rule 110 - Wolfram—Alpha (https://www.wolframalpha.com/input/?i= rule+110)
[13] Cook M 2004 Universality in elementary cellular automata *Complex Sys.* **15** 1
[14] ConwayLife, Glider Synthesis (https://conwaylife.com/wiki/Glider_synthesis)
[15] Adamatzky A 2010 *Game of Life Cellular Automata* (Berlin: Springer)
[16] Berlekamp E R, Conway J H and Guy R K 2004 *Winning Ways for Your Mathematical Plays* (London: Taylor and Francis)
[17] Adamatzky A (ed) 2001 *Collision-Based Computing* (Berlin: Springer)
[18] Gärtner B and Zehmakan A N 2001 Color war: cellular automata with majority-rule *Language and Automata Theory and Applications: Proc. 11th Int. Conf., Umeå* (Lecture Notes in Computer Science vol 10168) ed F Drewes, C Martín-Vide and T Bianca (Cham: Springer)
[19] Land M and Belew R K 1995 No perfect two-state cellular automata for density classification exists *Phys. Rev. Lett.* **74** 5148
[20] Busić A, Fatès N, Mairesse J and Marcovici I 2013 Density classification on infinite lattices and trees *Electron. J. Probab.* **18** 1
[21] LifeWiki (https://www.conwaylife.com/wiki/Main_Page)
[22] Scholarpedia (http://www.scholarpedia.org/article/Game_of_Life)
[23] Wezorek J A gallery of cellular automata created via artificial selection using Lifelike (http://jwezorek.com/CA/ca_gallery.html)
[24] Ingram S, Lee J and Arjunakani A hexagonal game of life (https://arunarjunakani.github.io/HexagonalGameOfLife/)
[25] Wolfram S 1985 Twenty problems in the theory of cellular automata *Phys. Scr.* **T9** 170
[26] Chaitin G J 1970 To a mathematical definition of 'life' *ACM SICACT News* **4** 12–18
[27] Marchal P 1998 John von Neumann: the founding father of artificial life *Artif. Life* **4** 229
[28] Bays C 1987 Candidates for the game of life in three dimensions *Complex Syst.* **1** 373
[29] Bays C 2006 A note about the discovery of many new rules for the game of three-dimensional life *Complex Syst.* **16** 381

Chapter 5

Brownian dynamics

Microscopic objects in liquid or gas environments are subject to continuous collisions with surrounding molecules (figure 5.1). This results in random motion, known as *Brownian motion* (also referred to as a *random walk* by mathematicians and as *diffusion* by physicists). One of the first experimental observations of Brownian motion was performed by Robert Brown (hence the name), a botanist who observed that microscopic particles within pollen grains were moving in random directions that were independent of each other. In 1905, Albert Einstein theoretically formulated the diffusion of small particles in liquid environments,

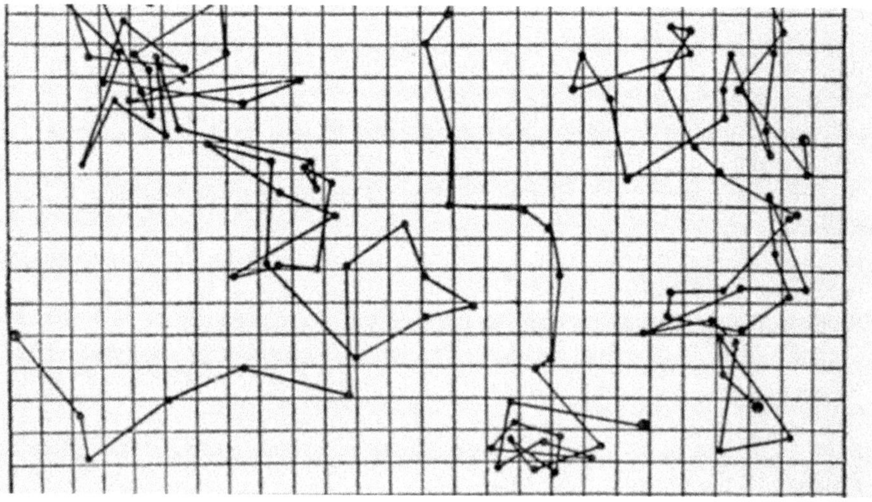

Figure 5.1. Brownian motion. In 1908, Jean Perrin recorded the individual trajectories of small (radius 0.53 μm) putty particles in water at thirty-second intervals. In the three trajectories shown above, successive positions are joined by straight-line segments. The observed random walk is due to collisions with water molecules. The mesh size is 3.2 μm. Reproduced from [1] with permission of EDP Sciences.

doi:10.1088/978-0-7503-3843-1ch5

connecting it to thermodynamics and the sizes of molecules [2]. Understanding the microscopic origin of Brownian motion has led to important physical insights into microscopic thermodynamics, nanotransport, and biological physics [3]. Furthermore, Brownian motion has been used to describe and explain phenomena in many different systems, such as animal foraging, financial markets, and human organizations [4].

Brownian motion can be described using *stochastic differential equations*, which are differential equations with an additional random (stochastic) driving term. To simulate such equations using finite differences, as we have done for molecular dynamics (see chapter 1), it is essential to learn how to simulate these random terms.

In this chapter, we start by simulating random walks with different step distributions and show that they belong to the same *universality class*, i.e., they share the same long-term macroscopic properties, even though their microscopic details are different. We then describe how to rescale random numbers with finite time differences. Next, we simulate Brownian motion in a liquid environment, highlighting its various timescales and when it is possible to neglect the inertia of Brownian particles, leading to important computational advantages. Finally, we simulate the motion of a Brownian particle held in an optical trap (i.e., a harmonic potential generated by a tightly focused laser beam) and analyze its statistical properties using multiple techniques.

Example codes: Example Python scripts related to this chapter can be found at: https://github.com/softmatterlab/SOCS/tree/main/Chapter_05_Brownian_Dynamics. Readers are welcome to participate in the discussions related to this chapter at: https://github.com/softmatterlab/SOCS/discussions/14.

5.1 Random walks and universality

A *random walk* is the motion of an agent that moves in a random direction at each time step. Let us start by considering the simplest case: in one dimension, an agent takes a step to the right with a probability of 50% and a step to the left with a probability of 50%. The corresponding iterative equation for this motion can be written as follows:

$$x_{i+1} = x_i + w_{\text{flip}}, \tag{5.1}$$

where w_{flip} represents a random selection between -1 and +1 (similar to the result of flipping a coin). We can identify several properties of this random walk. First, this random walk is symmetric, which means that $\langle x_i \rangle = 0$ for all x_i (assuming the initial condition is $x_0 = 0$, and indicating with $\langle \dots \rangle$ an ensemble average). Therefore, the mean displacement $\langle x_i \rangle$ is not a very informative measure for the motion of this particle. Instead, we can measure the mean square displacement (MSD) of the particle, which can be written as $\langle x_i^2 \rangle$. We can easily compute this for a few x_i's:

$$\langle x_1^2 \rangle = \frac{1}{2}(-1)^2 + \frac{1}{2}(+1)^2 = 1$$

$$\langle x_2^2 \rangle = \frac{1}{4}(-2)^2 + \frac{1}{2}(0)^2 + \frac{1}{4}(+2)^2 = 2$$

$$\langle x_3^2 \rangle = \frac{1}{8}(-3)^2 + \frac{3}{8}(-1)^2 + \frac{3}{8}(+1)^2 + \frac{1}{8}(+3)^2 = 3.$$

In fact, it is not difficult to show that $\langle x_i^2 \rangle = i$ (start by deriving $\langle x_{i+1}^2 \rangle = \langle x_i^2 \rangle + 1$, and then use induction).

It is possible to generate other kinds of random walk, with different motion rules (can you think of some?). Although the microscopic details of the motion that occurs at each time step can be different, the macroscopic outcome is remarkably similar for many different underlying motions. In the first exercise, we are going to check that for scales at which individual steps are not distinguishable, all random walks look the same, which is an example of *universality*.

Exercise 5.1. Universality of random walks. Write a program that simulates random walks with the following microscopic motion rules:

- $+1$ or -1 with equal probability (like a coin flip; see figure 5.2(a))
- a Gaussian distribution with a mean of zero and a variance of one (figure 5.2(b))
- equal probabilities of being $[-1, (1-\sqrt{3})/2, (1+\sqrt{3})/2]$ (figure 5.2(c)) (what are the variance and mean of a step in this case?).

a. Plot some sample trajectories for each case (see figure 5.2(d)–(f)). Note how the resulting trajectories are macroscopically indistinguishable, despite their microscopically different steps.

b. Plot the distributions of x_{1000} for each case and show that all three cases result in the same spread of the agent (see figure 5.2(g)–(i)).

5.2 Discrete white noise

We are now going to consider a random walker that, at each time step, moves by a Gaussian-distributed distance (in a similar way to the second case of exercise 5.1). In fact, this is a discrete version of a *Wiener process*, which is yet another name used by mathematicians to refer to Brownian motion. The Wiener process is the solution of the following stochastic differential equation:

$$\dot{x}(t) = W(t), \tag{5.2}$$

where $W(t)$ is a *white noise* (a quite complex mathematical object that is often employed in signal analysis). Equation (5.2) is arguably the simplest version of a free diffusion equation.

In a numerical simulation, equation (5.2) can be discretized as:

$$x_{i+1} = x_i + W_i \Delta t,$$

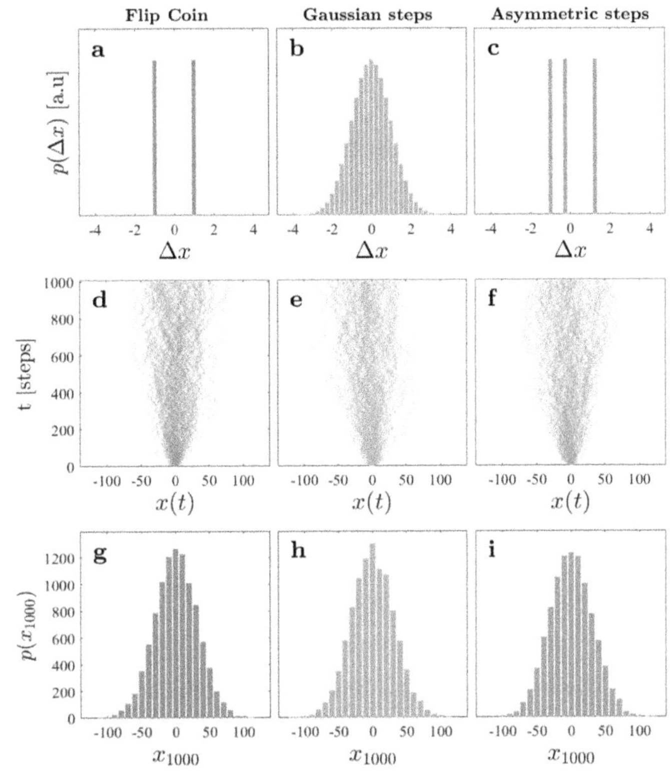

Figure 5.2. Universality of random walks. (a)–(c) The distributions of single steps for (a) a coin-flip step, (b) a Gaussian step, and (c) an asymmetric step. All the steps have the same mean ($\langle \Delta x \rangle = 0$) and variance ($\langle \Delta x^2 \rangle = 1$). (d)–(f) Resulting sample trajectories. (g)–(i) Distributions of the final positions of the agents after 1000 steps. As can be seen by comparing the three cases, the trajectories and distributions look indistinguishable on a macroscopic scale, i.e., when sufficiently many steps have been taken (can you think of any way of distinguishing them?).

where W_i is a random number extracted from a Gaussian distribution with a zero mean and a variance of σ_W^2. It is easy to see that $\langle x_n \rangle = 0$ because the individual steps have a zero mean, which is in agreement with what we would expect (why?).

If we now calculate the MSD, we obtain:

$$\langle x_n^2 \rangle = n\sigma_W^2 \Delta t^2 = \sigma_W^2 \Delta t \ t,$$

where $t = n\Delta t$. We now notice something strange: this MSD depends on our choice of the time step Δt! This is clearly not acceptable in a simulation of a physical system, as the simulation parameters should not have an effect on reality. In order to make our results independent of our choice of Δt, we need to rescale W_i as we change Δt (i.e., so that $\sigma_W^2 \Delta t$ is constant). This can be achieved by setting

$$W_i = \frac{w_i}{\sqrt{\Delta t}},$$

where w_i is a random number with a Gaussian distribution with a mean of zero and a variance of one.

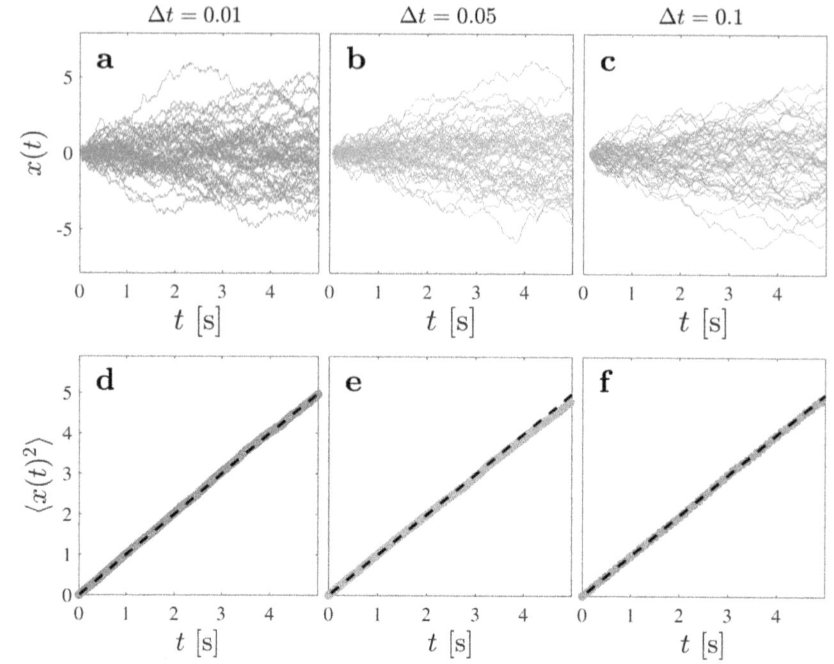

Figure 5.3. Time scaling of the diffusion equation with different time steps. (a)–(c) Sample (50) trajectories simulating the diffusion equation (5.3) for (a) $\Delta t = 0.01$ s, (b) $\Delta t = 0.05$ s, and (c) $\Delta t = 0.1$ s. Observe that the trajectories behave very similarly, despite the different Δt. (d)–(f) The statistical properties of these trajectories can be quantified by calculating their mean square displacements (averaged here over 10^4 realizations).

In conclusion, the numerical difference equation for diffusion with white noise can be written as:

$$x(t + \Delta t) = x(t) + w_i\sqrt{\Delta t}. \tag{5.3}$$

This equation permits simulations to obtain consistent results with different time steps, as we will numerically verify in the next exercise. Normally, the time step Δt should be much smaller than the characteristic timescales of the stochastic process to be simulated. If Δt is comparable to or larger than the smallest timescale, the numerical solution will not converge and will typically show a nonphysical oscillatory or divergent behavior. The case of free diffusion treated in this section is special because there is no characteristic timescale, as can be seen from the fact that equation (5.3) is self-similar under a rescaling of time (can you verify this?), and therefore, there is no optimal choice of Δt (i.e., any value of Δt works equally well).

Exercise 5.2. Discrete white noise. Simulate the diffusion equation (5.3) for $\Delta t = 0.01$, $\Delta t = 0.05$, and $\Delta t = 0.1$.

a. Obtain 50 different five-second trajectories for each Δt (see figure 5.3(a)–(c)). Observe that they qualitatively look very similar.

b. Simulate 10^4 trajectories for each Δt and calculate the average MSD for each ensemble. Plot MSD vs t for all three (see figures 5.3(d)–(f)). Observe that the results do not depend on our choice of Δt.

5.3 Brownian motion

One of the simplest examples of a stochastic system is a Brownian particle, which is a microscopic particle suspended in a fluid. Such a particle performs a random walk in its environment. A free Brownian particle undergoes a diffusion process that is very similar to what we described in the previous section. In fact, Brownian particles are often used to study random phenomena, because their motion due to thermal agitation from collisions with the surrounding fluid molecules provides a well-defined random background behavior that is dependent on the temperature and the fluid viscosity.

The motion of a Brownian particle in one dimension can be modeled by the following Langevin equation:

$$m\ddot{x} = -\gamma\dot{x} + \sqrt{2k_{\mathrm{B}}T\gamma}\ W(t), \tag{5.4}$$

where x is the particle position, m is its mass, γ is its friction coefficient, $\sqrt{2k_{\mathrm{B}}T\gamma}\ W(t)$ is the fluctuating force due to random impulses from the many neighboring fluid molecules, k_{B} is Boltzmann's constant, and T is the absolute temperature. Equation (5.4) can be solved numerically by considering the corresponding finite difference equation

$$m\frac{x_i - 2x_{i-1} + x_{i-2}}{(\Delta t)^2} = -\gamma\frac{x_i - x_{i-1}}{\Delta t} + \sqrt{2k_{\mathrm{B}}T\gamma}\frac{1}{\sqrt{\Delta t}}w_i \tag{5.5}$$

whose solution for x_i is

$$x_i = \frac{2 + \Delta t(\gamma/m)}{1 + \Delta t(\gamma/m)}x_{i-1} - \frac{1}{1 + \Delta t\frac{\gamma}{m}}x_{i-2} + \frac{\sqrt{2k_{\mathrm{B}}T\gamma}}{m\left[1 + \Delta t\frac{\gamma}{m}\right]}(\Delta t)^{3/2}w_i. \tag{5.6}$$

The ratio $\tau = m/\gamma$ is the momentum relaxation time, i.e., the timescale of the transition from smooth ballistic behavior to diffusive behavior. The time τ is very small, typically of the order of a few nanoseconds [3, 5]. (Note that this is very often orders of magnitude smaller than the timescales of typical experiments. Only since 2010 has it been possible to experimentally measure a particle's position sufficiently fast to probe its instantaneous velocity and its transition from the ballistic to the diffusive regime [6]. This was possible for a particle in air, for which the viscosity is around 200 times smaller than that for the same particle in water.) This is because the mass term in equation (5.5) becomes much smaller than the other terms when Δt is large. If we ignore that term by setting $m = 0$, the solution for x_i becomes:

$$x_i = x_{i-1} + \sqrt{\frac{2k_{\mathrm{B}}T\Delta t}{\gamma}}\,w_i. \tag{5.7}$$

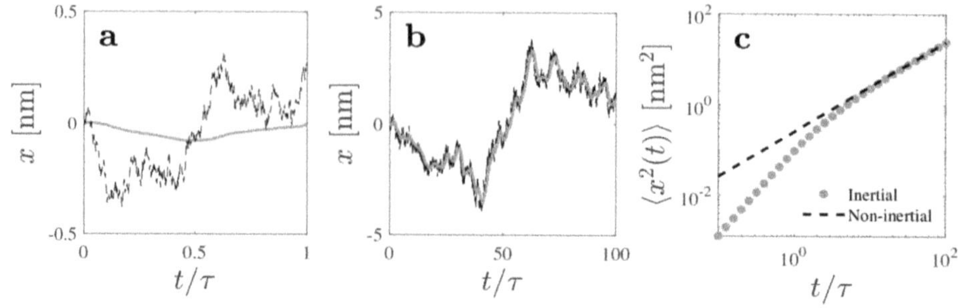

Figure 5.4. Brownian motion in inertial and viscous regimes. (a) Trajectories of a Brownian particle described by the Langevin equation with inertia (equation (5.6), black dashed line) and the Langevin equation without inertia (equation (5.7), cyan solid line). When the timescales are smaller than or comparable to τ (0.588 microseconds for this case), the difference is evident. (b) Trajectories simulated for much longer times than τ. In this case, the two trajectories overlap as the microscopic details become less prominent. (c) Mean square displacement measurements of the solutions with (black dashed line) and without inertia (cyan dots). The statistical properties of the two trajectories converge for $t \gtrsim 10\tau$. The parameters used for these simulations are: $m = 1.11 \times 10^{-14}$ kg, $T = 300$ K, $R = 1\ \mu$m, $\eta = 0.001$ Ns^{-1}, $\gamma = 6\pi\eta R$.

If we compare this to the original difference equation with mass (equation (5.6)), equation (5.7) is a very good approximation to Brownian motion for long time steps (i.e., for $\Delta t \gg \tau$), but it shows clear deviations at short timescales (i.e., $\Delta t \approx \tau$ or smaller). We will observe these differences in the next exercise.

Exercise 5.3. Brownian motion in inertial and viscous regimes. Consider a spherical Brownian particle with a radius $R = 1\ \mu$m and a mass $m = 1.11 \cdot 10^{-14}$ kg, immersed in water ($\eta = 0.001$ Ns^{-1}, $\gamma = 6\pi\eta R$, $T = 300$ K) that performs free diffusion.

a. Simulate the Langevin equation with mass (equation (5.6)) and without mass (equation (5.7)). Set $\Delta t < 0.1\tau$ and simulate trajectories that are least 100τ long using both methods. First, plot trajectories for a duration of τ and observe their differences (see figure 5.4(a)). Next, plot trajectories for a duration of 100τ and observe their differences (see figure 5.4(b)).

b. Compute the MSDs of both trajectories (by averaging several results). Show that the MSDs from the two approximations become similar for timescales that are large compared to τ, as shown in figure 5.4(c).

c. Compute time-averaged MSDs (averaged over a single, very long trajectory) and ensemble-averaged MSDs (averaged over many realizations). Show that they are equivalent to each other. This shows that the motion of a Brownian particle is *ergodic*.

5.4 Optical tweezers

In many cases, it is very useful to trap Brownian particles at a certain place to study their structure or behavior. This can be realized by a standard optical tweezers, which traps microscopic particles near the focal area of a highly focused laser beam [7]. This is particularly useful, for example, to study the physical or chemical properties of cells, to characterize the properties of colloidal particles, and to analyze single molecules or atoms. Since its discovery, optical trapping has revolutionized research many fields, such as cell biology, biophysics, nanotechnology, and single molecule physics; as a result, its inventor, Arthur Ashkin, received the Nobel Prize in Physics in 2018 [8].

A Brownian particle that is held in an optical trap can be modeled by a Langevin equation with a restoring force:

$$m\ddot{x} = -kx - \gamma\dot{x} + \sqrt{2k_\mathrm{B}T\gamma}\,W(t), \tag{5.8}$$

where k is the stiffness of the optical trapping potential. In most experimental scenarios, the timescales of the measurements are much longer than the characteristic velocity relaxation time (τ), as we have seen in the previous section. Therefore, in most cases, we can safely ignore the inertial term when Δt is significantly larger than microseconds. In this case, the solution for x_i for an optically trapped particle becomes:

$$x_i = x_{i-1} - \frac{k}{\gamma}x_{i-1}\Delta t + \sqrt{\frac{2k_\mathrm{B}T\Delta t}{\gamma}}\,w_i. \tag{5.9}$$

Note that although the diffusion equation is indefinitely rescalable to larger time steps, equation (5.9) has a relaxation time for the position that scales with the characteristic time of the optical trap (γ/k). This means that the time steps we choose when simulating a Brownian particle in an optical trap should be significantly smaller than this characteristic time ($\Delta t \ll \gamma/k$).

Several important properties can be computed from the trajectory of a Brownian particle in an optical trap:

- **Probability distribution:** At thermodynamic equilibrium, the probability distribution of a Brownian particle in a potential $U(x)$ obeys a *Boltzmann distribution*:

$$p(x) \propto \exp\left(-\frac{U(x)}{k_\mathrm{B}T}\right), \tag{5.10}$$

  where $U(x) = \frac{1}{2}kx^2$ in a harmonic trapping potential.
- **Positional autocorrelation function:** $C_x(t) = \langle x(t+t')x(t')\rangle$ indicates how long it takes for a particle to forget its current location. In a harmonic trapping potential, we have

$$C_x(t) = \frac{k_\mathrm{B}T}{k}\exp\left(-\frac{kt}{\gamma}\right). \tag{5.11}$$

- **Positional variance:** $\sigma^2(x) = \langle x^2\rangle$ indicates how much the particle position deviates from the center of the trap during its motion. In a harmonic trapping potential, we have

$$\sigma^2(x) = \frac{k_\mathrm{B}T}{k}. \tag{5.12}$$

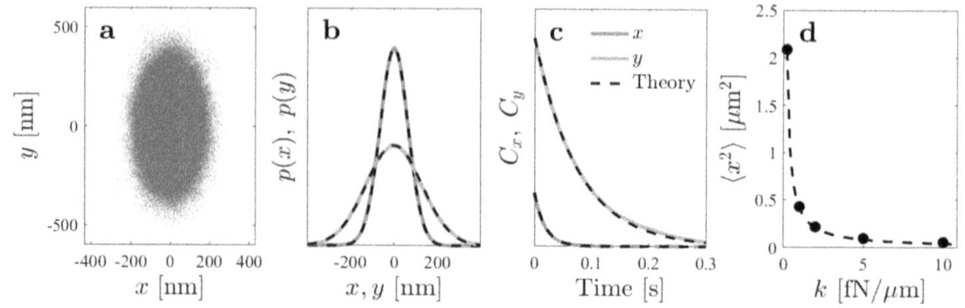

Figure 5.5. Brownian motion of an optically trapped particle. (a) 2D trajectory of a Brownian particle held in a harmonic potential with stiffness values of $k_x = 1$ pN μm^{-1} and $k_y = 0.25$ pN μm^{-1}. The particle moves further away from the center of the trap along the y-axis, as the stiffness in the y direction is weaker. (b) Numerically obtained probability distributions of the particle positions as a function of x (cyan line) and y (green line). These agree with the respective Boltzmann distributions (equation (5.10), black dashed lines). (c) Autocorrelation functions of the particle trajectory for x (C_x, cyan line) and y (C_y, green line). They match their theoretical predictions (equation (5.11), black dashed lines). (d) Variance of the particle position in one dimension as a function of the trapping stiffness k. These findings also match the results obtained using the theoretical formula (equation (5.12), black dashed line).

Exercise 5.4. Optically trapped Brownian particle. Consider a spherical Brownian particle with a radius $R = 1\ \mu$m immersed in water ($\eta = 0.001$ Ns^{-1}, $\gamma = 6\pi\eta R$, $T = 300$ K) that is held in an optical trap with different stiffness values in the x and y directions ($k_x = 1$ pN μm^{-1}, $k_y = 0.25$ pN μm^{-1}).

a. Simulate the motion of the particle in both dimensions using equation (5.9) and obtain trajectories for x and y. Plot the 2D trajectory of the particle (see figure 5.5(a)).

b. Calculate the probability distributions $p(x)$ and $p(y)$ of the particle and show that they follow the Boltzmann distribution (see figure 5.5(b)). Can you also compute the 2D probability distribution and compare it to its theoretical expected value?

c. Compute the autocorrelation functions C_x and C_y, and show that they fit the theoretical prediction (see figure 5.5(c)).

d. Simulate the particle motion in 1D for a range of stiffness values and plot the variance of the particle as a function of the trapping stiffness. Show that the particle position variance is inversely proportional to the stiffness (see figure 5.5(d)).

5.5 Further reading

This chapter is inspired by reference [9]. Detailed explanations and deeper theoretical derivations on the topics of Brownian motion, optical trapping, related experimental techniques, and data analysis methods can be found in reference [7]. Comprehensive reviews of the latest advances in the field of optical tweezers and their applications, as well as the various research fields that optical trapping has advanced, can be found in references [10, 11].

Reference [6] describes the first measurement of the instantaneous velocity of a Brownian particle, which confirmed several theoretical thermodynamic predictions. At the time, such measurements were made possible by high-frequency positional analysis combined with a low-viscosity environment (a Brownian particle trapped in air). With more recent advances in data acquisition speeds, direct measurement of the velocity of Brownian particles has also become possible in liquid environments [12].

To understand the principles of optical trapping and how it works in detail, a detailed numerical presentation of optical trapping forces using ray optics can be found in reference [13]. A software toolbox is provided for the direct calculation of the optical forces produced by ray optics. This permits one to also compute the optical trapping forces and torques exerted on non-spherical Brownian particles by focused laser beams.

For a generalization to non-spherically-symmetric particles moving in two or more dimensions, one must use the concept of a diffusion matrix in order to also include rotational degrees of freedom. An important case is the diffusion matrix of an ellipsoid, which was calculated by Perrin [14, 15]. A clear, self-contained reference for the diffusion matrix formalism is given by [16]. For a quick guide to the implementation of rotational dynamics, one can refer to reference [17].

Reference [18] provides an excellent mathematical perspective on the trajectories, fluctuations, and statistics of the stock market values, which have very strong similarities to those of Brownian motion.

5.6 Problems

Problem 5.1. Colloidal molecules. Consider a colloidal molecule that consists of two identical Brownian microspheres ($R = 0.2\,\mu$m) connected by a rigid nanorod of length $L = 1\,\mu$m of negligible thickness (the nanorod keeps the distance between the microspheres constant).

- **a.** Simulate this system undergoing free diffusion in 2D. Make an animation that displays the ensuing motion. *[Hint: At each time step, let both particles freely diffuse and correct their distance.]*
- **b.** Calculate the orientation of the rod φ at each time step and plot this as a function of time. *[Hint: Unwrap φ to account for its periodicity of 2π.]*
- **c.** Compute the angular mean square displacement $\langle\varphi^2(t)\rangle$ as a function of time for different values of L and observe how the rotational diffusion of this colloidal molecule is proportional to L^{-3}.

Problem 5.2. Kramers' transitions. Consider a colloidal particle that is held in a double-well potential: $U(x) = -ax^3 + bx$. Such a particle will have two stable equilibrium locations that are placed symmetrically at positions $\pm d$ around the origin and separated by an energy barrier E_b.

- **a.** Calculate a and b analytically in terms of d and E_b.
- **b.** Simulate the trajectory of the particle for various parameters. Plot the trajectory and observe that the particle has a bistable motion. Verify that the particle obeys the Boltzmann distribution.

c. Show that the frequency of the jumps over the potential barrier decays exponentially with increasing barrier height, i.e., $f \propto \exp\left(-\frac{E_b}{k_B T}\right)$, where f stands for the average frequency of the jumps.

Problem 5.3. Stochastic resonant damping. Consider a Brownian particle that is subject to an optical trapping potential whose center oscillates with a frequency of f and an amplitude of A:

$$F(x(t), t) = -k[x(t) - A\sin(2\pi f t)].$$

a. Simulate the motion of this particle and explore what happens with different values of f and A.

b. Show that, if the values of f and k/γ are comparable, the variance of the particle position does not monotonically decrease as k increases, so that there is an optimum value of k that achieves the minimum variance of the particle position. *[Hint: This phenomenon is known as stochastic resonant damping, see reference* [19]. *]*

Problem 5.4. Stochastic resonance. Consider a Brownian particle that is in a double-well potential that is periodically tilted:

$$F(x(t), t) = -ax^3 + bx + c\sin(2\pi f t),$$

where a, b, and c are constant, and f is the tilting frequency. If f is comparable to the particle's average jump frequency between the wells, there can be a partial synchronization of the jumps with the oscillating force—a phenomenon known as *stochastic resonance* (see reference [20]). This synchronization strongly depends on the temperature of the system.

a. Simulate such a Brownian particle. Observe what happens with different parameters and plot the resulting trajectories.

b. Compute the correlations of the particle location with the potential oscillations $C = \langle x(t)\sin(2\pi f t)\rangle$ and plot it as a function of temperature.

c. Show that, for fixed a, b, and c, there is an optimum frequency f which minimizes the particle's average potential energy inside this double-well potential.

Problem 5.5. Underdamped optical trapping. When the friction coefficient γ is sufficiently low (e.g., for a Brownian particle in air at low pressure), we cannot ignore the mass of the Brownian particle when simulating its motion.

a. Derive the difference equation of a Brownian particle that is optically trapped in air.

b. Show that for low values of γ, the autocorrelation function of the particle position (C_x) features an oscillatory behavior, while it gradually transitions to the exponential decay shown in figure 5.4(c) as γ increases.

5.7 Challenges

Challenge 5.1. Janus particle in 3D. A typical Janus particle is a microscopic spherical glass particle half-coated with a thin gold layer. It is a spherically-symmetric object from the point of view of diffusion; however, it is easy to observe its change of orientation. Simulate the overdamped Brownian dynamics of a Janus particle in three dimensions, also taking into account its rotational degrees of freedom.

Challenge 5.2. Ellipsoid in 3D. Simulate the overdamped Brownian dynamics of a prolate ellipsoidal particle in three dimensions, while also taking into account its rotational degrees of freedom. *[Hint: The diffusion coefficient in the direction of the longer semi-axis is different from the diffusion coefficient in the perpendicular direction. Refer to Perrin's expression of the diffusion coefficients* [14, 15]*.]*

Challenge 5.3. Chiral particles in 3D. 'Fusilli' are a traditional type of pasta made in the shape of a spiral. This geometrical shape is chiral, i.e., if we take its mirror image, it cannot be superposed onto the original shape by any rigid motion in three-dimensional space. Chiral shapes have an intrinsic coupling between the translational and rotational degrees of freedom along their principal axes. Simulate the overdamped Brownian dynamics of a spiral particle in three dimensions, while also taking into account its rotational degrees of freedom. Measure the coupling between the translational and rotational degrees of freedom. *[Hint: It might be useful to calculate the diffusion tensor of the chiral particle using Hydro++* [21, 22]*.]*

Challenge 5.4. Rigid body in 3D with inertia. In analogy to equation (5.4) and the relative derivation of the finite difference scheme, write the differential equation for the equation of motion of a rigid body in three dimensions, including the rotational degrees of freedom. From the resulting system of equations, derive a finite difference scheme that simulates the full 3D dynamics. *[Hint: Write the rotational equations of motion with respect to the center of mass of the particle.]*

References

[1] Perrin J 1910 Mouvement Brownien et molécules *J. Phys. Theor. Appl.* **9** 5–39

[2] Einstein A 1905 On the motion of small particles suspended in liquids at rest required by the molecular-kinetic theory of heat *Ann. Phys.* **17** 208

[3] Nelson E 2020 *Dynamical Theories of Brownian Motion* vol 106 (Princeton, NJ: Princeton University Press)

[4] Thompson J M T and Stewart H B 2002 *Nonlinear Dynamics and Chaos* (New York: Wiley)

[5] Purcell E M 1977 Life at Low Reynolds Number *Am. J. Phys.* **45** 3

[6] Li T, Kheifets S, Medellin D and Raizen M G 2010 Measurement of the instantaneous velocity of a Brownian particle *Science* **328** 1673–5

[7] Jones P H,Ieragò O M and Volpe G 2015 *Optical Tweezers: Principles and Applications* (Cambridge: Cambridge University Press)

[8] Kumar G R 2018 Nobel Prize in Physics: a gripping and extremely exciting tale of light *Curr. Sci.* **115** 1844–8

[9] Volpe G, Gigan S and Volpe G 2014 Simulation of the active Brownian motion of a microswimmer *Am. J. Phys.* **82** 659

[10] Polimeno P *et al* 2018 Optical tweezers and their applications *J. Quant. Spectrosc. Radiat. Transf.* **218** 131–50

[11] Gieseler J *et al* 2021 Optical tweezers—from calibration to applications: a tutorial *Adv. Opt. Photon.* **13** 74–241

[12] Kheifets S, Simha A, Melin K, Li T and Raizen M G 2014 Brownian motion in liquids at short times: instantaneous velocity and memory loss *Science* **343** 1493–6

[13] Callegari A, Mijalkov M, Gököz A B and Volpe G 2015 Computational toolbox for optical tweezers in geometrical optics *JOSA* B **32** B11–9

[14] Perrin F 1934 Mouvement Brownien d'un ellipsoide (I). Dispersion díelectrique pour des molécules ellipsoidales *J. Phys. Radium* **5** 497–511

[15] Perrin F 1936 Mouvement Brownien d'un ellipsoide (II). Rotation libre et dépolarisation des fluorescences. Translation et diffusion de mollécules ellipsoidales *J. Phys. Radium* **7** 1–11

[16] Fernandes M X 2002 Brownian dynamics simulation of rigid particles of arbitrary shape in external fields *Biophys. J.* **83** 3039–48

[17] Callegari A and Volpe G 2019 Numerical simulations of active Brownian particles *Flowing Matter (Soft and Biological Matter)* ed F Toschi and M Sega (Cham: Springer) pp 211–38

[18] Osborne M F M 1959 Brownian motion in the stock market *Oper. Res.* **7** 145–73

[19] Volpe G, Perrone S, Rubi J M and Petrov D 2008 Stochastic resonant damping in a noisy monostable system: theory and experiment *Phys. Rev.* E **77** 051107

[20] Gammaitoni L 1998 Stochastic resonance *Rev. Mod. Phys.* **70** 223–87

[21] García de la Torre J, del Rio Echenique G and Ortega A Hydro++ (http://leonardo.inf.um.es/macromol/programs/hydro++/hydro++.htm)

[22] García de la Torre J, del Rio Echenique G and Ortega A 2007 Improved calculation of rotational diffusion and intrinsic viscosity of bead models for macromolecules and nanoparticles *J. Phys. Chem.* B **111** 955–61

Chapter 6

Anomalous diffusion

The discovery of Brownian motion was groundbreaking for our understanding of the motion and behavior of microscopic particles, foraging animals, and stock markets. However, only a small fraction of real-world examples can be perfectly described by the *regular diffusion* equation of Brownian motion (see also chapter 5). From biology to finance, many commonly observed systems perform random motion of a sort different from diffusive Brownian motion. This general class of random walks is known as *anomalous diffusion*. Anomalous diffusion underlies various physical and biological systems, such as the motion of microscopic particles

Figure 6.1. Foraging birds. The motion of birds seeking food is an example of anomalous diffusion. These cranes perform irregular flights to search for food. While their motion is still random, their flights make the process persistent and superdiffusive. Picture captured by Aykut Argun near the lake Hornborgasjön, Sweden.

doi:10.1088/978-0-7503-3843-1ch6

in a crowded subcellular environment, the active dynamics of biomolecules in cytoplasm, and the foraging behavior of some animals [1–4].

By contrast with those of standard diffusive motion, the trajectories of anomalous diffusion scale nonlinearly with time. Their mean square displacement is nonlinear, i.e., MSD $\propto t^{\alpha}$ and they have an *anomalous diffusion exponent* $\alpha \neq 1$. The condition $\alpha < 1$ refers to *subdiffusion*, in which, for example, the movement of organelles is hindered because of a crowded cellular environment. The condition $\alpha > 1$ refers to *superdiffusion*, in which, for example, foraging animals actively move while looking for food (figure 6.1).

Anomalous diffusion can be generated by different theoretical models. The most commonly employed ones are: the *continuous-time random walk* (CTRW), *fractional Brownian motion* (FBM), the *Lévy walk* (LW), *annealed transient time motion* (ATTM), and *scaled Brownian motion* (SBM). Different models can be observed in different real-world systems.

In this chapter, we describe anomalous diffusion and its *ergodicity*. We start by demonstrating that a Brownian particle has a superdiffusive diffusion exponent ($\alpha > 1$) at inertial timescales and a subdiffusive exponent ($\alpha < 1$) if it is trapped. We then introduce regularization techniques for stochastic trajectories that are sampled irregularly in time. Next, we introduce different models that underlie different types of anomalous diffusion motion, namely, SBM, CTRW, and LW. Finally, we show that Brownian particles feature different anomalous diffusion behavior in non-homogeneous force fields, e.g., speckle light fields.

Example codes: Example Python scripts related to this chapter can be found at: https://github.com/softmatterlab/SOCS/tree/main/Chapter_06_Anomalous_Diffusion. Readers are welcome to participate in the discussions related to this chapter on: https://github.com/softmatterlab/SOCS/discussions/15.

6.1 Anomalous diffusion exponent

Given that the trajectories of a particle undergoing anomalous diffusion are intrinsically stochastic, it is necessary to use statistical averages to characterize their motion. As we have already seen in chapter 5, the displacement of a diffusive particle from its initial position can be characterized using the mean square displacement (MSD):

$$\mathrm{MSD}(t) = \langle [x(t + \tau) - x(\tau)]^2 \rangle, \tag{6.1}$$

where $\langle \cdot \rangle$ represents the average. This average can be calculated by taking the average of many different realizations, i.e., an *ensemble average*. To compute this, we run the same process n times and use the following formula:

$$\mathrm{eMSD}(t) = \frac{1}{n} \sum_i [x_i(t + \tau) - x_i(\tau)]^2, \tag{6.2}$$

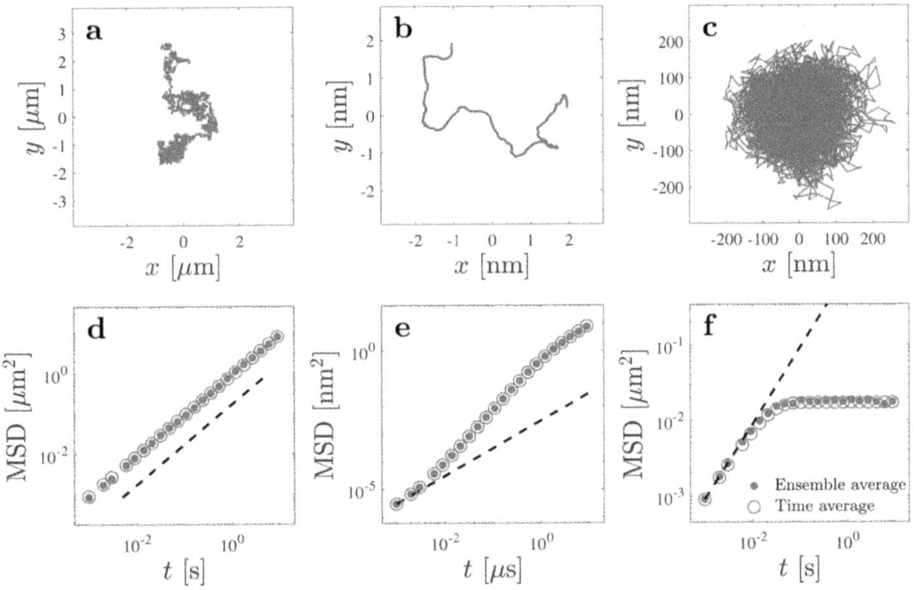

Figure 6.2. Diffusion regimes for a Brownian particle. Trajectories of (a) an overdamped Brownian particle (10 s), (b) an inertial Brownian particle (10 μs), and (c) a trapped Brownian particle (10 s). (d)–(f) Ensemble-averaged (equation (6.2), dots) and time-averaged (equation (6.3), circles) MSDs, which are diffusive ($\alpha = 1$) for the free Brownian particle, superdiffusive ($\alpha > 1$) for the inertial Brownian particle, and subdiffusive ($\alpha < 1$) for the trapped Brownian particle. The black dashed lines represent the slope corresponding to $\alpha = 1$. For all cases, the MSDs obtained using the ensemble average and the time average are equal, which shows that the motion of a Brownian particle is ergodic under all these conditions.

where x_i is a simulation run of the same diffusive process. Alternatively, the MSD can be calculated as a *time average*, i.e., an average over different values of the initial time τ. In order to compute this, instead of having an ensemble of short trajectories, we use a single longer trajectory and the following formula:

$$\mathrm{tMSD}(t) = \frac{1}{n}\sum_i [x(t + \tau_i) - x(\tau_i)]^2, \tag{6.3}$$

where τ_i is a reference time at which to calculate the square displacements of the particle.

For standard diffusive motion, the ensemble-averaged MSD calculated using equation (6.2) and the time-averaged MSD calculated using equation (6.3) are equal. Stochastic systems that have this statistical property are called *ergodic*. We will show in the next section that most anomalous diffusion models are *non-ergodic*.

For example, we can consider some Brownian particles, such as those we learnt about in chapter 5. For the trajectory of a freely diffusing overdamped Brownian particle (figure 6.2(a)), the MSD scales linearly with time (MSD $\propto t$, figure 6.2(d)). The MSD of an inertial Brownian particle at short timescales (figure 6.2(b)) is superdiffusive, due to its ballistic motion at short timescales (MSD $\propto t^\alpha$ with $\alpha > 1$, figure 6.2(e)). The MSD of a trapped Brownian particle (figure 6.2(c)) is subdiffusive (MSD $\propto t^{\alpha}$ with $\alpha < 1$, figure 6.2(f)) because the trap limits the motion of the particle.

Exercise 6.1. Diffusion regimes for a Brownian particle. Simulate the trajectories of a Brownian particle under different conditions, including an overdamped Brownian particle (as shown in figure 6.2(a)), an inertial Brownian particle (as shown in figure 6.2(b)), and a trapped Brownian particle (as shown in figure 6.2(c)). *[Hint: Refer to equations (5.7), (5.6) and (5.9), respectively.]*

a. Calculate the MSD of the overdamped Brownian particle using both the ensemble average (eMSD, equation (6.2)) and the time average (tMSD, equation (6.3)). Show that the exponent of the MSD is one, which confirms that this is regular diffusion. Also, show that the eMSD and the tMSD are equal, which confirms that this motion is ergodic.

b. Calculate the eMSD and the tMSD of the inertial Brownian particle. Show that the exponent of the MSD is greater than one, which confirms that the motion is superdiffusive. Verify that the eMSD and the tMSD are equal.

c. Calculate the eMSD and the tMSD of the trapped Brownian particle. Show that the exponent of the MSD is smaller than one, which confirms that the motion is subdiffusive. Verify that the eMSD and the tMSD are equal.

6.2 Regularization and normalization

Anomalous diffusion can be modeled using different mathematical algorithms, which we introduce in the next section. Most of these models produce trajectories that are sampled non-uniformly in time. Unfortunately, a non-uniformly-sampled trajectory is not easy to work with, e.g., to calculate its statistical properties, such as the MSD. Therefore, to be able to compare the characteristics of different anomalous diffusion trajectories, we need to be able to work around the irregular sampling of these trajectories.

Assume that we have an irregularly sampled trajectory (τ_j, u_j). A first-order algorithm that *regularizes* this trajectory goes through the following steps:

1. Create a regular sequence of times t_i that increase linearly from t_1 to t_{end} by equal increments of δt.
2. For each time step t_i, find $\tau_{\text{prev}} = \max(\tau_j \leqslant t_i)$ and $\tau_{\text{next}} = \min(\tau_j > t_i)$ and the corresponding u_{prev} and u_{next}.
3. Calculate each x_i such that $x_i = u_{\text{prev}} + \frac{u_{\text{next}} - u_{\text{prev}}}{\tau_{\text{next}} - \tau_{\text{prev}}}(t_i - \tau_{\text{prev}})$.
4. Return the regularized trajectory (t_i, x_i).

Of course, one can employ higher-order algorithms. However, this often fails to lead to more accurate results for diffusive trajectories (why?).

In addition to regularizing, it is often necessary to *normalize* the trajectories. In fact, to be able to compare and characterize different diffusion models, it is important to rescale the single trajectories so that they match in scale. This can be done using the following steps:

1. Take the increments of the trajectory $\Delta x_i = x_i - x_{i-1}$.
2. Normalize the increments in such a way that their standard deviation is one.
3. Take the cumulative sum of the increments to recover the normalized trajectory.
4. Subtract the mean from the trajectory.

This normalization protocol only rescales the amplitude of the trajectory without altering any temporal statistical information or correlations that are contained in the trajectory—in particular, it does not alter the anomalous exponent (why?).

Exercise 6.2. Regularizing an irregularly sampled trajectory.

a. Create a random walk trajectory with times whose increments are randomly distributed, as shown as by the blue circles in figure 6.3(a).

b. Regularize this trajectory as explained in this section. Plot both trajectories together to show that they follow the same path (orange crosses, figure 6.3(a)).

c. Write a function to obtain a normalized trajectory (orange line, figure 6.3(b)) from a generic trajectory (blue line, figure 6.3(b)) in such a way that its mean is zero and the standard deviation of its increments is one.

6.3 Models of anomalous diffusion

Here, we will describe the details of some algorithms that model anomalous diffusion.

Scaled Brownian motion (SBM)

An SBM consists of a standard Brownian motion (see chapter 5) with a deterministically varying diffusion coefficient [5]. This is achieved by rescaling the time after generating a standard Brownian motion trajectory (hence the name). A standard diffusive trajectory has an MSD that is proportional to t, i.e., MSD $\propto t$. Therefore, by rescaling the time to obey $t_s = t^{1/\alpha}$, we obtain an MSD that has a different exponent, i.e., MSD $\propto t_s^{\alpha}$. In more detail, to simulate the trajectory of an SBM with an anomalous diffusion exponent of α, we do the following:

1. Simulate a regular Brownian motion trajectory and generate a series of positions x_s corresponding to times t_s.
2. Rescale the array of times such that $t_s = t^{1/\alpha}$.
3. Regularize the trajectory (as shown in figure 6.3(a)).

This model can generate anomalous diffusion trajectories that are both subdiffusive ($\alpha < 1$) and superdiffusive ($\alpha > 1$). SBM is a non-ergodic model.

Exercise 6.3. Simulating scaled Brownian motion. In this exercise, we will simulate and analyze SBM trajectories.

a. Start by creating SBM trajectories with $\alpha = 0.5$, $\alpha = 1$, and $\alpha = 2$ using the algorithm described above. Normalize and plot each trajectory, as shown in figure 6.4(a).

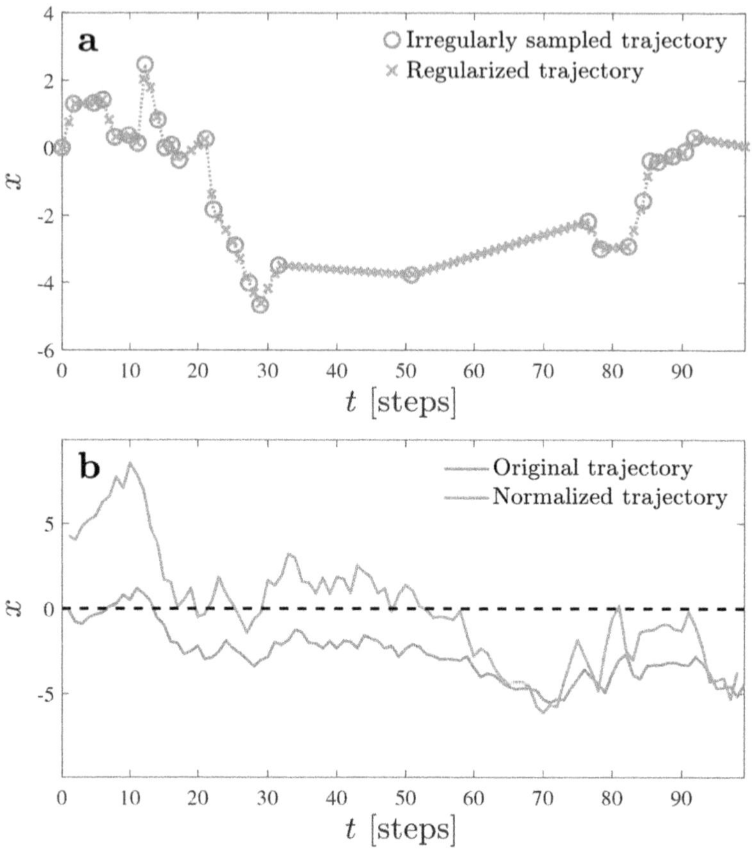

Figure 6.3. Regularization and normalization of a trajectory. (a) A trajectory that is irregularly sampled in time (blue circles) can be regularized by estimating its positions at uniformly sampled time steps (orange crosses). This permits us to compare differently sampled trajectories. (b) A trajectory can be normalized in such a way that its mean is zero and the standard deviation of its increments is one (blue line).

b. Simulate and plot 2D SBM trajectories for the same exponents, as shown in figure 6.4(b).

c. Calculate the eMSD (equation (6.2)) and tMSD (equation (6.3)) of an SBM trajectory for $\alpha = 1.5$. Show that these measurements do not match, as shown in figure 6.4(c). Conclude that SBM is non-ergodic. Why?

d. Change α and observe the differences between the eMSD and the tMSD; find the value of α for which the SBM trajectories become ergodic.

Continuous-time random walk

Like SBM, CTRW is also a modification of the standard Brownian motion [6]: at each step, the particle performs movements over Gaussian-distributed distances and

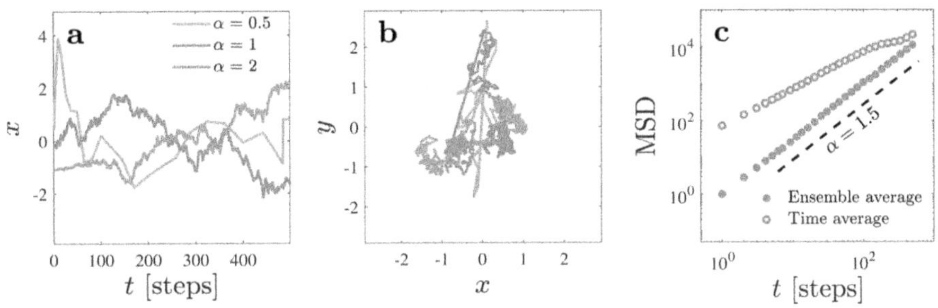

Figure 6.4. Scaled Brownian motion. (a) 1D and (b) 2D trajectories of SBM for $\alpha = 0.5$ (green lines), $\alpha = 1$ (blue lines), and $\alpha = 2$ (orange lines). (c) MSDs of the 1D SBM trajectories for $\alpha = 1.5$ (black dashed line) obtained using the ensemble average (dots) and the time average (circles). It is evident that the time average does not reproduce the anomalous diffusion exponent correctly, which shows that this model is non-ergodic.

then waits for a random time δt that has a power-law distribution that obeys $p(\delta t) \sim \delta t^{-1/\alpha}$ $(0 < \alpha < 1)$. A CTRW trajectory with an anomalous diffusion exponent α can be simulated using the following steps:

1. Simulate a regular Brownian motion trajectory.
2. Generate the waiting times between jumps $\delta t_i = r_i^{(-1/\alpha)}$, where r_i is a uniform random number between zero and one.
3. Calculate the corresponding time array as the cumulative sum of δt_i, i.e., $t_i = \sum_{k=1}^{i} \delta t_k$.
4. Regularize the trajectory (as shown in figure 6.3).

This model can only generate anomalous diffusion trajectories that are sub-diffusive ($\alpha < 1$) or diffusive ($\alpha = 1$). Also, this model is non-ergodic.

Exercise 6.4. Simulating a continuous-time random walk. In this exercise, we will simulate and analyze CTRW trajectories.

a. Start by creating CTRW trajectories with $\alpha = 0.5$ and $\alpha = 1$ using the algorithm described above. Regularize and plot each trajectory, as shown in figure 6.5(a).

b. Simulate and plot 2D CTRW trajectories for the same exponents, as shown in figure 6.5(b).

c. Show that CTRW is non-ergodic. Calculate the eMSD (equation (6.2)) and the tMSD (equation (6.3)) of a CTRW trajectory for $\alpha = 0.5$. Show that these measurements do not match, as shown in figure 6.5(c).

Lévy walk (LW)

Unlike the previous two anomalous diffusion models, in an LW the particle moves with a constant velocity at each time step but in a random direction [7]. At each step, the particle keeps walking in the same direction for a random amount of time that

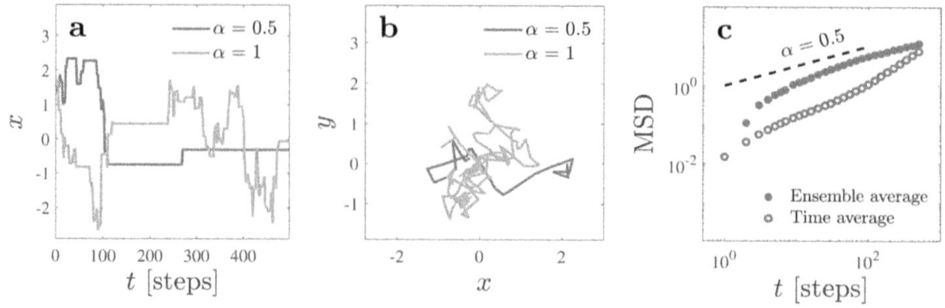

Figure 6.5. Continuous-time random walk. (a) 1D and (b) 2D trajectories of a CTRW for $\alpha = 0.5$ (blue lines) and $\alpha = 1$ (orange lines). (c) MSDs of the 1D CTRW trajectories for $\alpha = 0.5$ (black dashed line) obtained using the ensemble average (dots) and the time average (circles). It is evident that the time average does not reproduce the anomalous diffusion exponent correctly, which shows that this model is non-ergodic.

has a power-law distribution that follows $p(\delta t) \sim \delta t^{-1/(3-\alpha)}$ $(1 < \alpha < 2)$. An LW trajectory with an anomalous diffusion exponent α can be simulated using the following steps:

1. Choose a constant velocity v.
2. Generate the walking times $\delta t_i = r_i^{-1/(3-\alpha)}$, where r_i is a uniform random number between zero and one.
3. Calculate the positions $x_{i+1} = x_i + w_i v \delta t_i$, where w_i is a random number that represents the direction of the random movement (in 1D, $w_i \in [-1, 1]$). (In 2D, we choose a random movement angle.)
4. Calculate the corresponding times as the cumulative sum of δt, i.e., $t_i = \sum_{k=1}^{i} \delta t_k$.
5. Regularize the trajectory (as shown in figure 6.3).

This model can only generate anomalous diffusion trajectories that are super-diffusive ($\alpha > 1$) or diffusive ($\alpha = 1$).

Exercise 6.5. Simulating a Lévy walk. In this exercise, we will simulate and analyze LW trajectories.

a. Start by creating LW trajectories with $\alpha = 1$ and $\alpha = 2$ using the algorithm described above. Regularize and plot each trajectory, as shown in figure 6.6(a).

b. Simulate and plot 2D LW trajectories for the same exponents, as shown in figure 6.6(b).

c. Calculate the eMSD (equation (6.2)) and the tMSD (equation (6.3)) of an LW trajectory for $\alpha = 2$, as shown in figure 6.6(c). What about an LW trajectory with $\alpha = 1$?

6.4 Anomalous diffusion in a non-homogeneous force field

Anomalous diffusion can emerge from the interaction of Brownian particles with their environment, typically when the environment is non-homogeneous, such as a

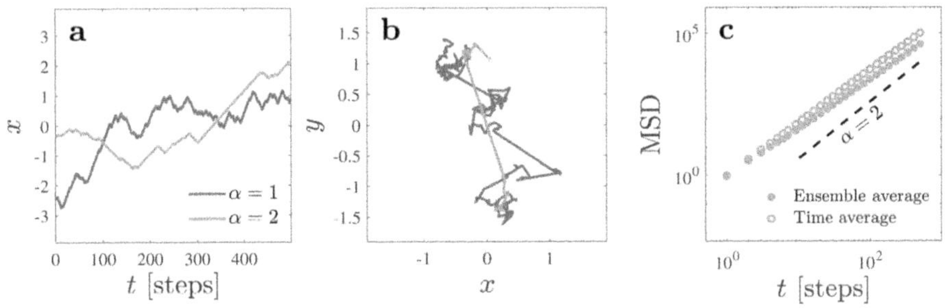

Figure 6.6. The Lévy walk model. (a) 1D and (b) 2D trajectories of Lévy walks (LWs) for $\alpha = 1$ (blue lines) and $\alpha = 2$ (orange lines). (c) MSDs of the LW trajectories for $\alpha = 2$ (black dashed line) obtained using the ensemble average (dots) and the time average (circles).

speckle light field [8, 9]. Such light fields naturally arise when a light source scatters through a rough medium, such as a biological tissue or a turbid liquid. Unlike optical tweezers, in which the Brownian particle is trapped in a single potential well, speckle light fields generate a random optical potential energy landscape with a number of local minima, between which the particle can diffuse.

In order to simulate a speckle field, we can generate randomly-located optical hotspots with power-law distributions of intensity and size. This can be achieved by adding randomly placed Gaussian intensity spots to an optical field. This creates an optical intensity field $I(x, y)$ whose derivative is proportional to the force field associated with the optical gradient force that attracts the particles toward the high-intensity regions. A Brownian particle in such a speckle light field is subject to forces that have an exponentially decaying correlation function. At very low power, these forces hardly affect the particle trajectory; therefore, a standard diffusion process is observed, as shown in figure 6.7(a). At intermediate power, these forces metastably trap the particle at the potential minima (figure 6.7(b)), which creates subdiffusive behavior over intermediate timescales, which transitions into regular diffusion ($\alpha = 1$) at longer timescales. At high power, the optical forces create some

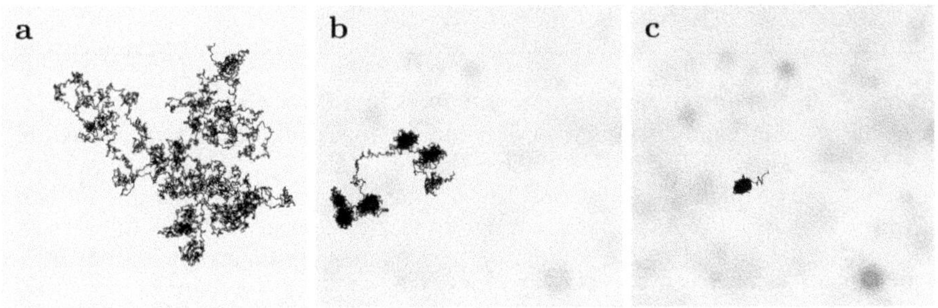

Figure 6.7. Anomalous diffusion of a Brownian particle in speckle light field. (a) When there is no speckle light field, the particle diffuses freely. (b) When the optical field has an intermediate strength, the particle becomes metastably trapped in different hotspots, but eventually keeps diffusing. (c) When the optical field is very strong, the particle becomes permanently trapped in a high-intensity region. The red-shaded background represents the intensity of the speckle light field and the black line the trajectory of a Brownian particle.

deep trapping potentials from which the particle cannot escape, as shown in figure 6.7(c).

Exercise 6.6. Subdiffusion in a speckle optical field. A Brownian particle (with a diffusion coefficient $D = 10^{-12}\ \mathrm{m^2 s^{-1}}$) placed in a speckle light field can be metastably or permanently trapped, depending on the intensity of the optical forces. We will create the speckle field by adding Gaussian hotspots $I(x, y) = I_0 \exp\left(-\frac{(x - x_c)^2 + (y - y_c)^2}{d^2}\right)$ to a 2D plane, where I_0 is the intensity, d is the decay length, and (x_c, y_c) is the position.

- **a.** Generate a speckle light field (with an area of $200\ \mu\mathrm{m} \times 200\ \mu\mathrm{m}$) with 200 Gaussian hotspots at random positions with a power-law-distributed decay length ($d(r_d) = d_0 r_d^{-1/2}$, $d_0 = 10\ \mu\mathrm{m}$) and a power-law-distributed intensity ($I(r_I) = I_0 r_I^{-1/2}$, $I_0 = 1$ mW), where r_d and r_I are independent uniform random numbers in [0, 1].
- **b.** Simulate the motion of a Brownian particle that is subject to gradient forces in this optical field (which drag the particle toward the hotspots with a velocity $V(x, y) = u\nabla I(x, y)$, where $u = 10^{-9}\ \mathrm{m^2 s^{-1} W^{-1}}$ is a constant related to the polarizability of the Brownian particle).
- **c.** Show that for small powers (the very low value of $I_0 = 0.1$ mW), the particle diffuses without being noticeably disturbed by the optical field, as shown in figure 6.7(a). Show also that for intermediate powers ($I_0 = 0.5$ mW), the particle becomes metastably trapped in different hotspots but eventually diffuses away, as shown in figure 6.7(b). Finally, show that for high powers (the very high value of $I_0 = 1$ mW), the particle becomes permanently trapped in one of the hotspots, as shown in figure 6.7(c). Play with the parameters and observe all these different outcomes.
- **d.** Calculate the MSDs for the different cases. Comment on the results.

The subdiffusion or trapping of a Brownian particle in a speckle light field depends on the polarizability of the particle [8]. Therefore, particles with different polarizabilities interact differently with the optical field. This can be used to sort or filter microscopic particles in a speckle field. In the next exercise, we will see an example of how this can be used.

Exercise 6.7. Sorting of microscopic particles in a flow. Consider a mixture of two different Brownian nanoparticles with different polarizabilities (e.g., particles A and B, for which the polarizability of B is double that of A, i.e., $u_A = u$ and $u_B = 2u$) flowing across a speckle light field. Consider a microfluidic channel with a size of $1000\ \mu\mathrm{m} \times 200\ \mu\mathrm{m}$. Create a microscopic speckle optical field with a size of $400\ \mu\mathrm{m} \times 200\ \mu\mathrm{m}$ placed in the middle of the channel. Assume that the particles are initially placed on the left-hand side and that the liquid flows at a speed of $V = 10\ \mu\mathrm{m}\ s^{-1}$ (dragging the particles). Simulate this system for many different values of the speckle field intensity and determine the average number of particles that remain trapped in the speckle light field. *[Hint: Check figure 4(a) of reference [9].]*

6.5 Further reading

Reference [4] provides a theoretical overview of different anomalous diffusion models and their mathematical properties. Reference [5] provides a detailed description of SBM. Reference [6] describes the CTRW model. For details of the Lévy walk model, refer to reference [7]. These papers provide an excellent theoretical foundation that explains why these models emerged, as well as a detailed analysis of the statistical properties of each anomalous diffusion model.

Two other important anomalous diffusion models that are not covered in this chapter are annealed transient time motion [10] (ATTM) and fractional Brownian motion (FBM) [11].

In 2020, a challenge was organized in order to characterize the anomalous diffusion of trajectories: the Anomalous Diffusion Challenge (AnDi Challenge: https://andi-challenge). This allowed scientists to compare the analytical and data--driven methods used to analyze these trajectories, which include inference, classification, and segmentation. Its results are given in reference [12]. Interestingly, several methods have been developed to analyze anomalous diffusion using advanced machine learning algorithms, which train neural networks (e.g., dense neural networks [13], recurrent neural networks [14], and graph neural networks [15]) to estimate the parameters and the underlying models of anomalous diffusion trajectories.

References [8, 9] show how Brownian particles can undergo subdiffusive or superdiffusive motion, depending on the dynamics of the speckle light field in which they are immersed.

6.6 Problems

Problem 6.1. Simulation of fractional Brownian motion. FBM is another anomalous diffusion model with correlated Gaussian increments [11]. Depending on the sign of this correlation, the output trajectories can become subdiffusive or superdiffusive.

a. Refer to reference [11] and simulate FBM for various values of α and compare them.

b. Compare the eMSD and the tMSD of FBM trajectories for various α values. Show that this model is ergodic.

Problem 6.2. Simulation of annealed transient time motion. Annealed transient time motion (ATTM) is another anomalous diffusion model that has random changes in the diffusion coefficient [10]. The waiting times between changes in the diffusion coefficient have a distribution of the form $p(\delta t) = \delta t^{\alpha+1}$. ATTM only models subdiffusive trajectories ($\alpha < 1$).

a. Refer to reference [10], simulate ATTM for various values of α, and compare the results.

b. Compare the eMSD and the tMSD of FBM trajectories for various α values. Show that this model is non-ergodic.

Problem 6.3. Subdiffusion in crowded environments. Simulate the motion of N Brownian particles with radius r. Assume that the particles are confined in a circular well with a radius of R. Show that the particles' motion becomes subdiffusive at long timescales. Study the MSD of these particles as a function of their packing fraction Nr^2/R^2.

Problem 6.4. Anomalous diffusion models with engineered environments. There can be an interesting interplay between anomalous diffusion and environmental complexity (e.g., due to the force fields of obstacles).

a. Examine the distributions of particles performing various kinds of anomalous diffusion in a circular well.

b. Determine the differences in the diffusion of particles with different underlying anomalous diffusion models and exponents in an infinite two-dimensional arena with periodic disc-shaped obstacles.

c. Engineer a force field to control the propagation of anomalously-diffusing particles.

Problem 6.5. Superdiffusion of a Brownian particle in a dynamical speckle light field. Design a speckle light field that is dynamic, which means that the positions and intensities of the hotspots change over time with a characteristic timescale τ.

a. Analyze the diffusion exponent of a particle that is immersed in this environment as a function of τ.

b. Show that, for slow changes in the light field, the particle motion is subdiffusive. However, if the change in the light field is fast, the particle's motion becomes superdiffusive.

6.7 Challenges

Challenge 6.1. Diffusive regimes of a free Brownian particle. Simulate the Brownian motion of a microscopic particle at timescales ranging from a fraction of nanosecond to milliseconds. Show the transition from diffusive to ballistic to diffusive motion that occurs while increasing the timescale.

Challenge 6.2. Characterization of anomalous diffusion trajectories. Simulate the anomalous diffusion trajectories of a random underlying model with different anomalous diffusion exponents. Use the eMSD to infer the anomalous diffusion exponents (α) of trajectories that have different lengths. Analyze the error in the inference as a function of the trajectory length. Try to program your own method that infers α more accurately than the eMSD method for short trajectories. Write a program that predicts the underlying model for an anomalous diffusion trajectory.

Challenge 6.3. Use of deep neural networks to characterize anomalous diffusion. Neural networks are a very powerful tool for the characterization of anomalous diffusion, i.e., to determine the anomalous diffusion exponent α, to classify the underlying anomalous diffusion model, and to segment trajectories that have transitions in their behaviors (see the task in the Anomalous Diffusion Challenge [12]). Write a program to characterize anomalous diffusion given a single trajectory using recurrent neural networks. Use simulated data to train your network. Compare your results against the methods that competed in the Anomalous Diffusion Challenge.

References

[1] Viswanathan G M, da Luz M G E, Raposo E P and Stanley H E 2011 *The Physics of Foraging: an Introduction to Random Searches and Biological Encounters* (Cambridge: Cambridge University Press)

[2] Barkai E, Garini Y and Metzler R 2012 Strange kinetics of single molecules in living cells *Phys. Today* **65** 29–35

[3] Höfling F and Franosch T 2013 Anomalous transport in the crowded world of biological cells *Rep. Prog. Phys.* **76** 046602

[4] Metzler R, Jeon J-H, Cherstvy A G and Barkai E 2014 Anomalous diffusion models and their properties: non-stationarity, non-ergodicity, and ageing at the centenary of single particle tracking *Phys. Chem. Chem. Phys.* **16** 24128–64

[5] Lim S C and Muniandy S V 2002 Gaussian processes for modeling anomalous diffusion *Phys. Rev.* E **66** 021114

[6] Scher H and Montroll E W 1975 Anomalous transit-time dispersion in amorphous solids *Phys. Rev.* B **12** 2455–77

[7] Klafter J and Zumofen G 1994 Lévy statistics in a Hamiltonian system *Phys. Rev.* E **49** 4873–7

[8] Volpe G, Volpe G and Gigan S 2014 Brownian motion in a speckle light field: tunable anomalous diffusion and selective optical manipulation *Sci. Rep.* **4** 3936

[9] Volpe G, Kurz L, Callegari A, Volpe G and Gigan S 2014 Speckle optical tweezers: micromanipulation with random light fields *Opt. Express* **22** 18159–67

[10] Massignan P *et al* 2014 Nonergodic subdiffusion from Brownian motion in an inhomogeneous medium *Phys. Rev. Lett.* **112** 150603

[11] Mandelbrot B B and Van Ness J W 1968 Fractional Brownian motions, fractional noises and applications *SIAM Rev.* **10** 422–37

[12] Muñoz-Gil G *et al* 2021 Objective comparison of methods to decode anomalous diffusion *Nat. Commun.* **12** 6253

[13] Gentili A and Volpe G 2021 Characterization of anomalous diffusion classical statistics powered by deep learning (CONDOR) *J. Phys. A: Math. Theor.* **54** 314003

[14] Argun A, Volpe G and Bo S 2021 Classification, inference and segmentation of anomalous diffusion with recurrent neural networks *J. Phys. A: Math. Theor.* **54** 294003

[15] Verdier H *et al* 2021 Learning physical properties of anomalous random walks using graph neural networks *J. Phys. A: Math. Theor.* **54** 234001

IOP Publishing

Aykut Argun, Agnese Callegari and Giovanni Volpe

Chapter 7

Multiplicative noise

In common parlance, the term *noise* indicates a loud, annoying, unnecessary sound. Typical examples of noisy environments are busy marketplaces, congested roads, and natural events, such as storms or hurricanes: if we try to speak to a friend in such circumstances, we have to raise our voice louder than usual, and in some cases even shouting will not be enough for our words to be received and understood.

In a more general sense, noise is a disturbance that prevents or hinders a *signal* from being clearly detected. In fact, some degree of noise is an unavoidable aspect of any communication, detection, recording, or measure. For example, a picture taken in conditions of scarce luminosity (e.g., a photograph of an astronomical object, figure 7.1)

Figure 7.1. Different kinds of noise. Noise is found everywhere—for example, in an astronomical photo. There are different kinds of noise; each is connected with different aspects of the photographic digital process. The first, *dark-current noise*, is of thermal origin and is independent of the object being imaged. The second, *readout noise*, is connected to the characteristics of a single detecting unit (pixel). The third, *shot noise* or *Poissonian noise*, depends on the strength of the signal received from the astronomical object; this is a *multiplicative noise*, because it depends on the strength of the signal. *Source: picture of the Orion nebula, captured by Aykut Argun.*

doi:10.1088/978-0-7503-3843-1ch7

is affected by electronic noise. Such noise is obviously also present in conditions of high luminosity, but the signal intensity is much stronger and the effect of noise less visible. In fact, there are different kinds of noise. There is *dark-current noise*, which has a thermal origin that is independent of the luminosity. There is also *readout noise*, which is connected with the properties of the single pixels used as detectors and, therefore, is different for different pixels, but constant in time for any given pixel. Then, there is *shot noise*, which typically depends on the light intensity according to a Poissonian distribution (for this reason, it is often also referred to as *Poissonian noise*). This noise is an example of *multiplicative noise*, i.e., a noise whose strength depends on the intensity of the signal. Another example of a system with multiplicative noise is a Brownian particle subject to a Brownian noise whose intensity depends on the particle's position.

In this chapter, we consider the Brownian motion of a particle at thermodynamic equilibrium with its environment in the presence of multiplicative noise. First, we introduce a minimal model for the motion of a Brownian particle in a confined space (see also chapter 5). In particular, we observe that, if the particle is at thermodynamic equilibrium with its environment (and not subject to any external forces), its *Boltzmann distribution* is uniform. Then, we show that, when introducing a multiplicative noise, it is necessary to complement it with a *spurious drift* (or *noise-induced drift*) to recover the correct equilibrium distribution. We also highlight how this spurious drift is connected to the mathematical foundations of stochastic calculus with multiplicative noise, in particular to the *Itô* and *Stratonovich integrals*. Finally, we show how this method can be applied to the simulation of Brownian particles in diffusion gradients generated by the presence of nearby surfaces.

Example codes: Example Python scripts related to this chapter can be found at: https://github.com/softmatterlab/SOCS/tree/main/Chapter_07_Multiplicative_Noise. Readers are welcome to participate in the discussions related to this chapter at: https://github.com/softmatterlab/SOCS/discussions/16.

7.1 A minimal discrete-time model

The Brownian motion of a particle in one-dimensional subject to non-deterministic forces can be modeled by the stochastic differential equation

$$\mathrm{d}x = \sigma \mathrm{d}W, \tag{7.1}$$

where W is a Wiener process, i.e., a stochastic process whose increments $\mathrm{d}W$ are independent and normally distributed with an average of zero and a standard deviation of one, and where $\sigma = \sqrt{2D}$ and D is the Stokes-Einstein diffusion constant. In the case of constant σ, the finite-difference equation that allows use to build the trajectory associated with a given realization of the Wiener process W is

$$x_{n+1} = x_n + \sqrt{\sigma^2 \Delta t}\ W_n. \tag{7.2}$$

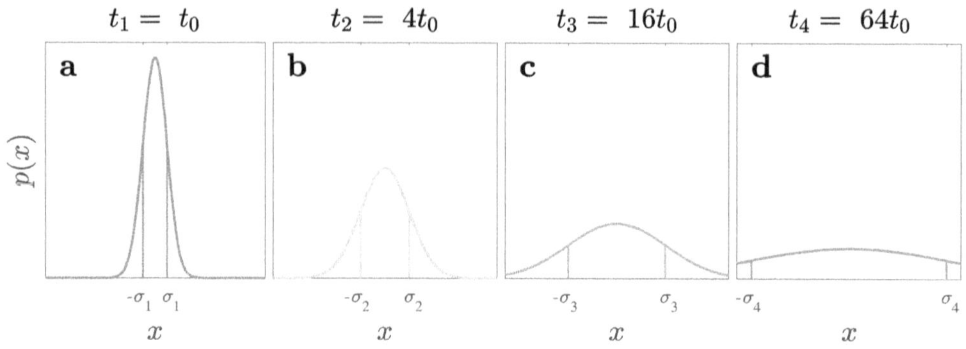

Figure 7.2. Free diffusion. The probability of finding a freely diffusing particle around its initial position $x(0) = x_0$ at times (a) $t_1 = t_0$, (b) $t_2 = 4\ t_0$, (c) $t_3 = 16\ t_0$, and (d) $t_4 = 64\ t_0$. The probability distribution is always a Gaussian centered at x_0 and with a standard deviation of $\sigma\sqrt{t}$ (i.e., (a) $\sigma_1 = \sigma$, (b) $\sigma_2 = 2\sigma$, (c) $\sigma_3 = 4\sigma$, and (d) $\sigma_4 = 8\sigma$). The probability of finding the particle within the interval $[x_0 - \sigma\sqrt{t}, x_0 + \sigma\sqrt{t}]$ is $\approx 68\%$. As time passes, the distribution becomes flatter and flatter.

(See chapter 5 for a detailed description of Brownian motion and its numerical solution).

There are an infinite number of different realizations of the Wiener process, and for each realization, there is a corresponding unique trajectory $x(t)$. Let us consider all these different realizations and write the probability distribution $p_t(x)$ of finding a particle at position x at time t. If the particle starts from x_0 at $t = 0$, the distribution is initially a delta function centered at $x = x_0$ and, for all $t > 0$, it can be shown that

$$p_t(x') \propto e^{-\frac{(x'-x_0)^2}{2\sigma^2 t}}. \tag{7.3}$$

The solution of this equation is shown in figure 7.2. At a time $t > 0$, the particle is still found more frequently in the proximity of its initial position within a characteristic distance $\sigma\sqrt{t}$. At any time, the probability of finding the particle outside the interval $[x_0 - \sigma\sqrt{t}, x_0 + \sigma\sqrt{t}]$ is smaller than 2 erf(1), i.e., smaller than 32%. As time passes, the probability distribution becomes flatter, i.e., the particle diffuses freely and is almost equally likely to be found at any given point around an ever-increasing interval around the initial position.

We can obtain an approximation of the distribution in equation (7.3)) by considering a sufficient number of independent realizations N of the trajectory $x(t)$ and recording their final point x_{fin} at a fixed time t_{fin}. This will be done in the following exercise.

Exercise 7.1. Free diffusion of a Brownian particle. Consider a particle initially located at $x = x_0$. Take, for simplicity, $\sigma = 1$ and $\Delta t = 1$ s. Generate $N = 10^4$ independent trajectories $x_j^{(n)}$ (n indicates the realization number and j the time step, $t_j = j\ \Delta t$). For all simulation runs, $x_0^{(n)} = x_0$. At each time step, the particle can move to the left or to the right with equal probability, so that

$$x_{j+1}^{(n)} = x_j^{(n)} \pm \sigma\sqrt{\Delta t}\,. \tag{7.4}$$

For various values of j, plot a histogram of $x_j^{(n)}$. Show that these are Gaussian distributions centered at x_0 and with a standard deviation of $\sigma\sqrt{2j\Delta t}$. Compare your results with figure 7.2.

The probability distribution of a particle diffusing on an infinite space (such as that in figure 7.2 and in the previous exercise) never reaches a steady state (why?). To overcome this, we now consider a particle diffusing within the bounded interval $[-L/2, L/2]$. If we place the particle at position $x(0) = 0$ at time $t = 0$, we expect that, after a sufficiently long time, the particle will acquire a flat distribution, i.e., it can be found anywhere in the interval with equal probability.

Exercise 7.2. Brownian particle in a box. Consider a particle confined by the interval $[-L/2, L/2]$ with $L = 100$ and moving according to equation (7.4). Following the notation used in exercise 7.1.1, take $\sigma = 1$, $\Delta t = 0.01$ s, and $x_0 = 0$. Implement *reflective boundary conditions*:

$$\begin{cases} x_j \to -L - x_j & \text{for} \quad x_j < -\dfrac{L}{2} \\ x_j \to x_j & \text{for} - \dfrac{L}{2} \leqslant x_j \leqslant \dfrac{L}{2} \\ x_j \to L - x_j & \text{for} \quad x_j > \dfrac{L}{2} \end{cases} \tag{7.5}$$

Simulate $N = 10^4$ independent trajectories for various durations (e.g., $T_{tot} = 10$, 10^2, 10^3, 10^4, and 10^5 s) and plot the histograms of the final positions. Compare your results with figures 7.3(a) and 7.3(b).

When completing exercise 7.2, we notice that, while, for a time $T_{tot} \lesssim 100$ s, one can still estimate the initial position of the particle by observing its probability distribution, for $T_{tot} \gtrsim 1000$ s we completely lose the ability to infer its initial position, because the probability distribution becomes flat (as it should be, since no external driving force is acting on the particle).

7.2 Position-dependent noise

Let us now examine the case of multiplicative noise, i.e., when the noise acting on the particle depends on the particle position. The equation of motion is:

$$\mathrm{d}x = \sigma(x)\,\mathrm{d}W. \tag{7.6}$$

We aim to answer the following question: How can we numerically simulate equation (7.6) using finite differences? In other words, can we use the same approach as with equation (7.1) (and chapter 5), or shall we employ some new tricks?

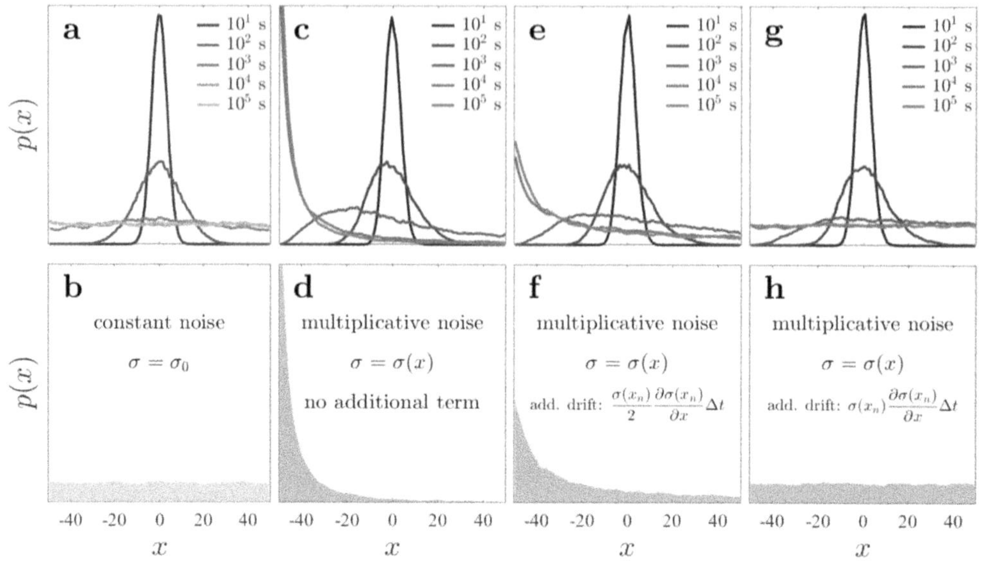

Figure 7.3. Brownian particle in a box. Numerical simulation of a Brownian particle in a box with (a)–(b) constant noise and (c)–(h) multiplicative noise. The top row shows the evolution of the simulated probability distribution after 10, 10^2, 10^3, 10^4, and 10^5 s. The bottom row highlights the probability distribution of the simulated steady state (reached after $\approx 10^5$ s). The simulations are performed according to (a)–(b) equation (7.4), (c)–(d) equation (7.7) (the *Itô integral*), (e)–(f) equation (7.11) with $\alpha = 0.5$ (the *Stratonovich integral*), and (g)–(h) equation (7.11) with $\alpha = 1$ (the *anti-Itô integral*). The results for multiplicative noise feature different steady-state distributions, depending on the simulation method. In thermodynamic equilibrium and in the absence of external forces acting on the particle, the steady-state probability distribution must be flat, as in (h).

Following the same procedure as for equation (7.1), we obtain a first guess for the finite-difference equation:

$$\text{(a guess!) } x_{j+1} = x_j \pm \sigma(x_j)\sqrt{\Delta t}, \tag{7.7}$$

where we calculate x_{j+1} using the value of the noise at x_j, i.e., $\sigma(x_j)$.

Exercise 7.3. Brownian particle with multiplicative noise. Building on the configuration of exercise 7.2 ($\Delta t = 0.01$ s, $x_0 = 0$, $L = 100$), we now introduce a multiplicative noise:

$$\sigma(x) = \sigma_0 + \frac{\Delta\sigma}{L}x, \tag{7.8}$$

where $x \in \left[-\frac{L}{2}, \frac{L}{2}\right]$, and σ_0 and $\Delta\sigma$ are such that $\sigma(x) > 0$ all over the domain (e.g., take $\sigma_0 = 1$ and $\Delta\sigma = 1.8$, so that at the extremes of the domain we have $\sigma\left(-\frac{L}{2}\right) = 0.1$ and $\sigma\left(-\frac{L}{2}\right) = 1.9$). In a similar way to exercise 7.2, we aim to build the probability distribution $p(x)$ of finding the particle at a given position of the interval—and we expect to find a flat distribution, because no deterministic external forces are acting on the particle.

a. Using equation (7.7), simulate $N = 10^4$ independent trajectories for various durations (e.g., $T_{\rm tot} = 10$, 10^2, 10^3, 10^4, and 10^5 s) and plot the histograms of the final positions. Compare your results with figures 7.3(c) and 7.3(d).

b. Why is the final distribution non-uniform? Is this compatible with a Brownian particle at thermodynamic equilibrium with its environment?

As figures 7.3(c) and 7.3(d) show, by using equation (7.7) to build the discrete trajectories x_j, we obtain a probability distribution that is not the one we would expect for the absence of external forces acting on a Brownian particle at thermodynamic equilibrium with its environment: in fact, the $p(x)$ obtained is not flat, but it has a peak near the left boundary where $\sigma(x)$ is at its smallest. It is not difficult to intuitively understand the reason for this outcome: close to the left boundary, the particle displacements are smaller in magnitude (and symmetrical in both directions), so that the particle lingers longer closer to that boundary (in this finite-difference model).

Despite being able to intuitively explain the outcome of the simulation, this is *not* what happens in physical systems *at thermodynamic equilibrium* in which no external force is at play. No external force entails a flat potential, which corresponds to a flat probability distribution. Therefore, equation (7.7) does *not* give the correct physical trajectory of the particle, and does *not* allow the correct equilibrium probability distribution of the particle in the confined system to be retrieved.

As we will see in the following sections, this problem can be fixed by introducing an extra term into equation (7.6), called *spurious drift*. However, to do this, we need to dive into a bit of the mathematical background beyond stochastic differential equations with multiplicative noise.

7.3 Stochastic integrals

To understand what is missing in equation (7.7), we have to step back to the meaning of the stochastic equation (7.1) and how it can be integrated. We can formally calculate the value of x at a time T as follows:

$$x(T) = \int_0^T f(x_t) \circ_\alpha \mathrm{d}W_t, \tag{7.9}$$

where

$$\int_0^T f(x_t) \circ_\alpha \mathrm{d}W_t \equiv \lim_{N\to\infty} \sum_{n=0}^{N-1} f\left(x_{t_n}^{(\alpha)}\right) \Delta W_{t_n}$$

with $t_n^{(\alpha)} = \frac{n+\alpha}{N}T$ and $\alpha \in [0, 1]$. Different choices of α might give different values for the stochastic integral (7.9), and therefore different realizations of $x_\alpha(t)$ (this is the reason why α appears in the symbol $\circ_\alpha$, which denotes the *integration convention*). In fact, a complete model is defined by the stochastic differential equation *and* the integration convention given by α.

We now try to give an intuitive meaning of what the value of α implies (the most common values are $\alpha = 0$, 0.5, and 1). When opting for a model with $\alpha = 0$ (the *Itô integral*), we calculate the value of $f(x)$ at the initial moment of each time interval. This is often the preferred choice in economics and biology. Importantly, this is particularly convenient in numerical simulations (can you see why?). When taking $\alpha = 0.5$ (the *Stratonovich integral*), we use the value of $f(x)$ at the middle of each time interval. This makes this choice particularly symmetric and a favorite for theoretical physicists (and also because the standard rules of calculus apply). Finally, taking $\alpha = 1$ (the *anti-Itô integral*), we evaluate $f(x)$ at the end of each time interval.

Conveniently, it is possible to transform any stochastic differential equation into the form with $\alpha = 0$ by introducing a *noise-induced drift* (or *spurious drift*):

$$\sigma(x_t)\circ_\alpha \mathrm{d}W_t = \underbrace{\alpha\sigma(x_t)\frac{\mathrm{d}\sigma(x_t)}{\mathrm{d}x}\mathrm{d}t}_{\text{noise-induced drift}} + \sigma(x)\,\mathrm{d}W_t, \tag{7.10}$$

where the stochastic integral convention for $\sigma(x)\mathrm{d}\,W_t$ is the Itô convention ($\alpha = 0$). This form is particularly convenient because it can be readily numerically integrated using the finite-difference approach we have already seen in chapter 5.

7.4 The spurious drift

Inspired by the previous section and the discussion about the different α integration conventions, we are now ready to check whether, by assuming an integration convention other than $\alpha = 0$ (which was implicitly used in exercise 7.3), we can retrieve the correct probability distribution $p(x)$. We are then going to integrate equation (7.10) with $\alpha = 0.5$ and $\alpha = 1$.

Exercise 7.4. Comparing stochastic integrals. Using the configuration and notation of exercise 7.3, we now explicitly introduce the convention we are using in the finite-difference equation, obtaining

$$x_{j+1} = x_j + \underbrace{\alpha\sigma(x_t)\frac{\mathrm{d}\sigma(x_t)}{\mathrm{d}x}\Delta t}_{\text{noise-induced drift}} \pm \sigma\sqrt{\Delta t}\,. \tag{7.11}$$

a. Set $\alpha = 0.5$ (the Stratonovich integral) and plot the probability distribution after $T_{\text{tot}} = 10$, 10^2, 10^3, 10^4, and 10^5 s. Compare your results with figure 7.3(e)–(f).

b. Set $\alpha = 1$ (the anti-Itô integral) and plot the probability distribution after $T_{\text{tot}} = 10$, 10^2, 10^3, 10^4, and 10^5 s. Compare your results with figure 7.3(g)–(h).

c. In the light of your results, which convention works best to reproduce the equilibrium distribution (i.e., the Boltzmann distribution)?

The integration with $\alpha = 1$ (the anti-Itô integral) is the one that allows us to obtain, after a sufficiently long time, the expected equilibrium probability distribution for a particle at thermodynamic equilibrium and not driven by any deterministic force. Therefore, when simulating the dynamics of systems that are affected by multiplicative noise, we must take into account the *spurious drift*:

$$v_{\text{spurious}} = \sigma(x_t)\frac{\mathrm{d}\sigma(x_t)}{\mathrm{d}x}. \tag{7.12}$$

Neglecting the spurious drift leads to the generation of biased, nonphysical trajectories.

Exercise 7.5. Multiplicative noise in a harmonic potential. Consider a particle in a harmonic potential with multiplicative noise.

a. Write down the equation of motion. *[Hint: You can start from equation (5.9). Also, very importantly, take into account the relation between diffusion and friction at thermodynamic equilibrium (equation (7.17)).]*

b. Simulate its motion according to the three stochastic integrals we have considered here.

c. Show that only using the anti-Itô integral ($\alpha = 1$) leads to the correct Boltzamnn distribution. *[Hint: See equation (5.10).]*

d. Show that the correlation function of the particle's motion in the presence of multiplicative noise is qualitatively different from that in its absence. *[Hint: See equation (5.11).]*

7.5 Drift and diffusion measurement

How can the diffusion coefficient and spurious drift be measured from a trajectory? Given a trajectory x_j with positions measured at equispaced time intervals $t_j = j\Delta t$, a partition of the space into bins centered at L_k with a width of ΔL, and a reference observation number of time steps n (corresponding to the observation time $\tau = n\Delta t$), we define the following quantities:

$$\text{drift: } C(L_k, \Delta L, n) = \frac{1}{n\Delta t}\langle x_{j+n} - x_j \rangle, \tag{7.13}$$

$$\text{biased diffusion: } D_{\text{b}}(L_k, \Delta L, n) = \frac{1}{2n\Delta t}\langle (x_{j+n} - x_j)^2 \rangle, \tag{7.14}$$

$$\text{unbiased diffusion: } D_{\text{u}}(L_k, \Delta L, n) = \frac{1}{2n\Delta t}\left\langle \left(x_{j+n} - x_j - C(L_j)\cdot n\Delta t\right)^2 \right\rangle, \tag{7.15}$$

where the average is taken over all x_j belonging to the interval $\left[L_k - \frac{\Delta L}{2}, L_k + \frac{\Delta L}{2}\right]$. It is possible to use different values of n for equations (7.13), (7.14), and (7.15). In particular, it might be convenient to use a very small n for the measurement of the diffusion, and a larger n for the calculation of the drift. In all cases, for an accurate estimate of the drift and the diffusion from a sequence of experimental data, a high sampling rate is to be preferred.

Exercise 7.6. Measurement of drift, diffusion, and spurious drift. Simulate long trajectories of a freely diffusing Brownian particle confined to a segment with constant diffusion (as in exercise 7.2) and with multiplicative noise (as in exercise 7.3).

a. In each case, calculate the drift using equation (7.13). Compare your results with figures 7.4(a) and 7.4(d).

b. Calculate the diffusion using equation (7.14). Compare your results with figures 7.4(b) and 7.4(e).

c. From the numerical diffusion, calculate the noise-induced drift. Compare your results with figures 7.4(c) and 7.4f.

d. Do you find any differences in the drifts calculated for trajectories without and with multiplicative noise?

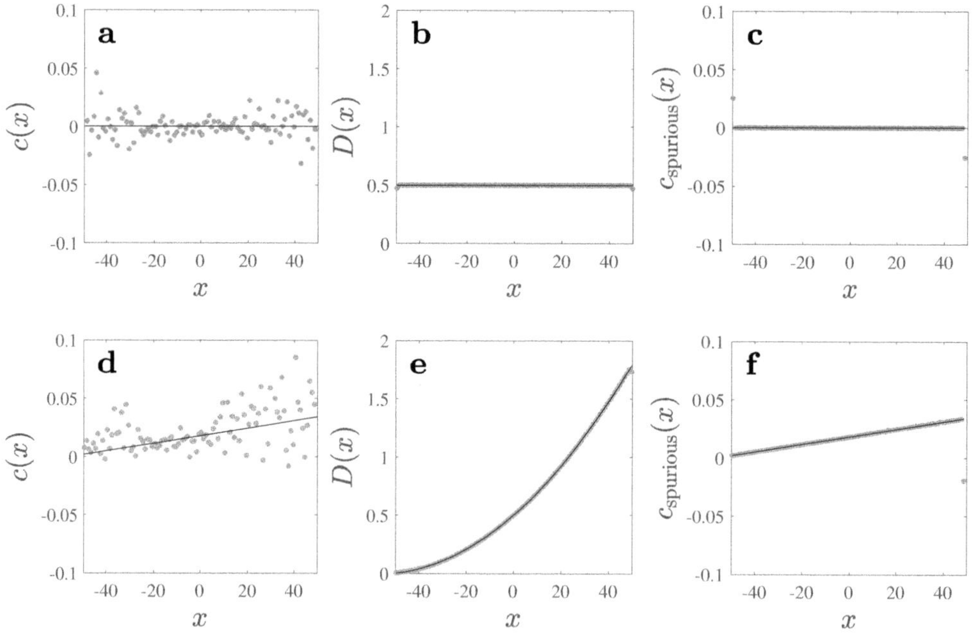

Figure 7.4. Measurement of drift, diffusion, and spurious drift. (a) Drift, (b) diffusion, and (c) spurious drift for a Brownian particle confined to a segment and subject to constant diffusion. (d) Drift, (e) diffusion, and (f) spurious drift for a Brownian particle confined to a segment with multiplicative noise. The black lines are theoretical, while the points are experimental. The noise-induced drift (c, f) is better calculated from the gradient of the diffusion (b, e) than directly from the trajectory (a, d) (why?). Why does the numerical value of the spurious drift calculated for the extremities differ significantly from the predicted values in (c) and (f)? *[Hint: think of the reflective boundary conditions and their effects on the trajectories at the boundary points.]* For the simulation, you can consider the parameters used in exercise 7.3. For the calculation of the drift and the diffusion you need to define a time lapse $\tau = 10\Delta t$.

In general, noise-induced drift, which depends on the spatial gradient of the diffusion, is smaller than the drift caused by a deterministic force. Therefore, it is commonly estimated by first calculating the diffusion and then calculating its gradient.

7.6 Particles close to an interface

In Nature and many technological applications, microscopic microorganisms or particles move in proximity to interfaces: examples include colloidal particles suspended above a substrate, bacteria swimming close to a liquid–air interface, and small molecules diffusing close to a cell membrane. In all these cases, the diffusion coefficient D (which is related to σ via $2D = \sigma^2$) determines the amount of thermal noise $\sigma \mathrm{d}W$. If D depends on the position x, then we have multiplicative noise, and the equation governing the dynamics is

$$\mathrm{d}x = \frac{F(x)}{\gamma(x)}\mathrm{d}t + \frac{\mathrm{d}D(x)}{\mathrm{d}x}\mathrm{d}t + \sqrt{2D(x)}\,\mathrm{d}W_t, \tag{7.16}$$

where $\frac{\mathrm{d}D(x)}{\mathrm{d}x}\mathrm{d}t$ is the noise-induced drift, $F(x)$ is an external force, and $\gamma(x)$ is the friction coefficient, which at thermodynamic equilibrium is related to the diffusion $D(x)$ via the Einstein relation

$$\gamma(x)\, D(x) = k_{\mathrm{B}}T, \tag{7.17}$$

where T is the absolute temperature and k_{B} is the Boltzmann constant ($k_{\mathrm{B}}T$ is the characteristic energy scale of the microscopic system).

Exercise 7.7. Perpendicular motion of a Brownian particle near an interface. Simulate the perpendicular motion of a Brownian particle near a flat surface (placed at $z = 0$, with reflective boundary conditions, figure 7.5(a)), accounting for spurious drift. The perpendicular diffusion coefficient of a particle near a boundary is given by $D_{\perp}(z) = \frac{D_{\mathrm{SE}}}{\xi_{\perp}(z)}$

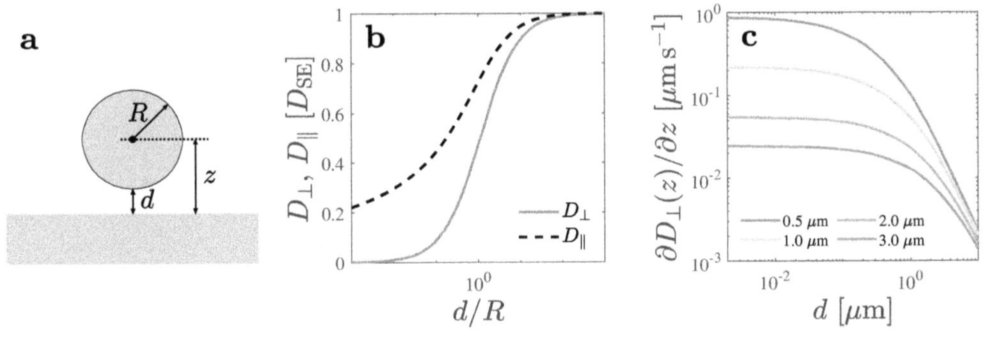

Figure 7.5. Diffusion coefficient of a particle near a planar interface. (a) Schematic of a spherical particle (cyan circle) with a radius of R near a planar interface (the cyan rectangle). Here, z is the center-to-surface distance, i.e., $z = R + d$, where d is the surface-to-surface distance. (b) Diffusion coefficients $D_{\perp}$ and $D_{\parallel}$ as a function of d/R, for no-slip boundary conditions. (c) Spurious drift calculated for spherical colloids of different radii, as a function of the surface-to-surface distance d: the spurious drift is more important for particles with smaller radii.

[1, 2], where $\xi_\perp(z) = \frac{4}{3}\sinh[a(z)]\sum_{n=1}^{\infty}\frac{n(n+1)}{(2n-1)(2n+3)}\left(\frac{2\sinh[(2n+1)a(z)] + (2n+1)\sinh[2a(z)]}{4\sinh^2[(n+0.5)a(z)] - (2n+1)^2\sinh^2[a(z)]} - 1\right)$ and $a(z) = \cosh^{-1}\left(1 + \frac{z}{R}\right)$, R is the radius of the particle, $D_{\mathrm{SE}} = \frac{k_{\mathrm{B}}T}{6\pi\eta R}$ is the Stokes-Einstein bulk diffusion coefficient, and η is the viscosity of the liquid.

a. Implement a function to numerically calculate $D_\perp(z)$ (the black dashed line in figure 7.5(b)) and its gradient $\frac{\mathrm{d}D_\perp(z)}{\mathrm{d}z}$ (figure 7.5(c)). Use this function to generate some free-diffusion trajectories of the particle.

b. Assuming the surface to be horizontal, add effective gravity (which induces a constant negative drift $-v_{\mathrm{g}}$) and a repulsive force that exponentially decays away from the surface (to prevent the particle from hitting the interface, e.g., $v_{\mathrm{r}}e^{-\frac{z-h}{\lambda}}$, which is a simple model for the electrostatic repulsion between a particle and a surface). Generate some trajectories and calculate the equilibrium distribution for the particle. *[Hint: It is important to choose physically meaningful values for v_{g}, v_{r}, h, and λ. We want the effective gravity to push the particle down toward $z = 0$, and the electrostatic repulsion to counteract the effective gravity in the close proximity of $z = 0$. The time step Δt should be sufficiently small. If with R we indicate the radius of the particle, then $h \approx R(1 + \alpha)$ with $0 < \alpha \ll 1$, $0 < \lambda \ll h - R$, v_{g} and v_{r} must be of the same order of magnitude. For example, $\alpha \approx 0.1$ and $\lambda \approx (h - R)/3$ should work. Δt must be such that $v_{\mathrm{g}}\Delta t$, $v_{\mathrm{r}}\Delta t$, and $v_{\mathrm{spurious}}\Delta t$ are much smaller than R.]*

If we want to simulate the full three-dimensional motion of the particle, we must include all the spatial dimensions. In the case of a spherically-symmetric particle, the stochastic differential equation reads:

$$\mathbf{dr} = \mathbf{dr}_{\mathrm{d}} + \mathbf{dr}_{\mathrm{s}}, \tag{7.18}$$

where $\mathbf{dr}_{\mathrm{d}}$ is the displacement caused by the deterministic forces and $\mathbf{dr}_{\mathrm{s}}$ is the stochastic term generated by the Brownian forces. In the case of a spherical particle near a planar interface, the deterministic displacement includes a term due to external forces and a term due to the noise-induced drift:

$$\mathbf{dr}_{\mathrm{d}} = \frac{1}{k_{\mathrm{B}}T}\mathbf{D}\,\mathbf{F}\,\mathrm{d}t + \nabla\cdot\mathbf{D}\,\mathrm{d}t, \tag{7.19}$$

where $\mathbf{D}$ is a diffusion matrix with the general form:

$$\mathbf{D} = \begin{bmatrix} D_{xx} & D_{xy} & D_{xz} \\ D_{yx} & D_{yy} & D_{yz} \\ D_{zx} & D_{zy} & D_{zz} \end{bmatrix} \tag{7.20}$$

where all terms depend on $\mathbf{r}$, and $\nabla\cdot\mathbf{D}$ is a vector with components $\nabla\cdot\mathbf{D}_x$, $\nabla\cdot\mathbf{D}_y$, $\nabla\cdot\mathbf{D}_z$ with

$$\nabla\cdot\mathbf{D}_x = \frac{\partial D_{xx}}{\partial x} + \frac{\partial D_{yx}}{\partial y} + \frac{\partial D_{zx}}{\partial z},\ \nabla\cdot\mathbf{D}_y = \frac{\partial D_{xy}}{\partial x} + \frac{\partial D_{yy}}{\partial y} + \frac{\partial D_{zy}}{\partial z},\ \text{and } \nabla\cdot\mathbf{D}_z = \frac{\partial D_{xz}}{\partial x} + \frac{\partial D_{yz}}{\partial y} + \frac{\partial D_{zz}}{\partial z}.$$

The deterministic force $\mathbf{F}(\mathbf{r})$ is a vector with components

$$\mathbf{F} = \begin{bmatrix} F_x \\ F_y \\ F_z \end{bmatrix}.$$

The Brownian displacement $d\mathbf{r}_s$ is a vector with components

$$d\mathbf{r}_s = \begin{bmatrix} \sqrt{2D_x(\mathbf{r})}\, dW_{x,t} \\ \sqrt{2D_y(\mathbf{r})}\, dW_{y,t} \\ \sqrt{2D_z(\mathbf{r})}\, dW_{z,t} \end{bmatrix}. \tag{7.21}$$

Here, z denotes the direction perpendicular to the planar interface, and x and y denote the two directions parallel to the interface.

In the case of a spherical particle, all the off-diagonal terms of the diffusion matrix (equation (7.20)) are zero. Furthermore, for symmetry reasons, $D_{xx} = D_{yy} \overset{def}{=} D_{||}$, where $D_{||}$ depends only on z. Moreover, $\nabla \cdot \mathbf{D}$ is non-zero only in its z component, which is equal to $\frac{\partial D_{\perp}}{\partial z}$. Thus, the equations of motion are:

$$\begin{cases} dx &= \dfrac{D_{||}(z)\, F_x(\mathbf{r})}{k_B T}\, dt + \sqrt{2D_{||}(z)}\, dW_{x,t} \\ dy &= \dfrac{D_{||}(z)\, F_y(\mathbf{r})}{k_B T}\, dt + \sqrt{2D_{||}(z)}\, dW_{y,t} \\ dz &= \dfrac{D_{\perp}(z)\, F_z(\mathbf{r})}{k_B T}\, dt + \dfrac{dD_{\perp}(z)}{dz} dt + \sqrt{2D_{\perp}(z)}\, dW_{x,t} \end{cases} \tag{7.22}$$

where $D_{\perp}(z)$ is as defined in exercise 7.7; we are going to see the definition of $D_{||}(z)$ in the next exercise.

Exercise 7.8. 3D motion of a Brownian particle near an interface. Simulate the full three-dimensional trajectory of a particle near a flat planar surface. $D_{\perp}(z)$ is defined in exercise 7.7. $D_{||}(z) = \frac{D_{SE}}{\xi_{||}(z)}$ [3], where

$$\xi_{||}(z) = \left\{1 + 0.498\left\{\log\left[1.027\left(\frac{R}{z - R}\right)^{0.986} + 1\right]\right\}^{1.027}\right\}^{0.979}.$$

a. Implement a function to numerically calculate $D_{||}(z)$ and its gradient $\frac{dD_{||}(z)}{dz}$.

b. Simulate the finite-difference equation for each degree of freedom and generate particle trajectories. Why are the equations for x and y free from spurious drift?

Exercise 7.9. Calculation of the noise-induced drift from the diffusion for a particle near a wall. Simulate several trajectories of the vertical position of a particle diffusing near a wall, as explained in exercise 7.7.

a. Calculate the drift and the diffusion from the generated trajectories.

b. Calculate the noise-induced drift by calculating the gradient of the diffusion. Do you expect to find that the noise-induced drift is substantially the drift? Or do you see other terms in the drift? *[Hint: Exercise 7.7 requested the addition of a gravitational drift and an electrostatic drift. Those are both deterministic. Do you expect to see them in the drift? How do they compare in magnitude with the noise-induced drift?]*

7.7 Further reading

A broad review of multiplicative noise in dynamical systems can be found in reference [4]. For an introduction to stochastic calculus and stochastic differential equations, we refer the reader to [5, 6].

For the case of a spherical particle near a planar surface, the main references for the diffusion coefficients are [1, 3]. References [7, 8] are insightful works on the physics of particles close to or at planar interfaces, focused on water–air interfaces (i.e., full-slip interfaces).

References [2, 9–11] are focused on experiments in which noise-induced drift plays an important role. They illustrate concrete cases in which noise-induced drift is measured in real systems and highlight the importance of a correct simulation scheme for the comparison of simulations with experiments.

Reference [12] connects the stochastic integral to the underlying microscopic dynamics of a system (out of thermodynamic equilibrium), showing also that the stochastic integral for a given system may vary with the operational conditions.

7.8 Problems

Problem 7.1. Shot noise in images. Numerically reproduce the effect of shot noise and explore its multiplicative nature. How is it different from dark-current noise? *[Hint: check reference [13] and the references therein.]*

Problem 7.2. Particle near a water–air interface. Simulate the dynamics of a spherical colloidal particle near a water–air interface, which is a full-slip interface. Then, determine the drift, diffusion, and noise-induced drift from the simulated trajectories. Discuss the feasibility of such an experiment and the requirements for the experimental acquisition equipment (e.g., the acquisition sampling rate to be used for the trajectory) and the material (e.g., the size of the particles to be used). *[Hint: for the dependence of $D_\perp$ and $D_\parallel$ for a full-slip interface, see reference [7].]*

Problem 7.3. Optically trapped particle near a flat surface. You need to optically trap a particle near the bottom slide of your sample (or near a water–air interface). The particle fluctuates around the trap center because of thermal forces. From a previous experiment, you know that you can trap the same particle in the bulk of the solution with an optical trap of power P_0 or higher. Focusing on the role played by the presence of a diffusion gradient, how does the minimum power required to trap the particle depend on the distance of the particle from the surface?

Problem 7.4. Particles optically trapped in the proximity of each other. To perform an experiment that reveals the interaction potential between two colloidal particles (for example, an experiment like the one described in reference [11]), you need to optically trap two particles close to each other (and far away from all other surfaces). Simulate the dynamics of the two optically trapped particles using the correct diffusion coefficient. *[Hint: for the diffusion coefficients between two nearby particles in the bulk, check reference [14].]*

Problem 7.5. Blinking optical tweezers. Read reference [11]. Simulate the dynamics of two particles in a blinking optical tweezers experiment, in which two particles are trapped in two optical traps for a certain amount of time T_{on}, released by switching off the optical traps for an amount of time T_{off}, and finally retrapped by switching the traps back on so that the cycle can be repeated. Choose appropriate values of T_{on} and T_{off} so that the particles can be released and retrapped with confidence. From the trajectories of the free evolution, retrieve the information of the drift and the diffusion coefficients of the two particles as a function of their distance.

Problem 7.6. Scaling of the spurious drift. Study how the magnitude of the spurious drift scales with the particle size.

7.9 Challenges

Challenge 7.1. Emergence of multiplicative noise. Demonstrate how multiplicative noise emerges from a molecular dynamics simulation of a Brownian particle close to a boundary, or of two Brownian particles close to each other in the bulk.

Challenge 7.2. Multiplicative noise in a temperature gradient. Study multiplicative noise and the spurious drift for a Brownian particle in a temperature gradient (e.g., by using molecular dynamics). Note that the temperature gradient effectively drives the particle out of thermodynamic equilibrium (What does this imply for the stochastic integral to be used?).

Challenge 7.3. Spurious drifts that matter. Can you think of some situations in which the presence of the spurious drift associated with multiplicative noise matters?

References

[1] Brenner H 1961 The slow motion of a sphere through a viscous fluid towards a plane surface *Chem. Eng. Sci.* **16** 242–51

[2] Brettschneider T, Volpe G, Helden L, Wehr J and Bechinger C 2011 Force measurement in the presence of Brownian noise: equilibrium-distribution method versus drift method *Phys. Rev.* E **83** 041113

[3] Nguyen A V and Evans G M 2004 Exact and global rational approximate expressions for resistance coefficients for a colloidal solid sphere moving in a quiescent liquid parallel to a slip gas–liquid interface *J. Colloid Interface Sci.* **273** 262–70
[4] Volpe G and Wehr J 2016 Effective drifts in dynamical systems with multiplicative noise: a review of recent progress *Rep. Prog. Phys.* **79** 053901
[5] Øksendal B 2003 *Stochastic Differential Equations: an Introduction with Applications* (Berlin: Springer)
[6] Karatzas I and Shreve S 2012 *Brownian Motion and Stochastic Calculus* (Berlin: Springer)
[7] Villa S 2018 Behaviour of a colloid close to an air-water interface: interactions and dynamics *PhD thesis* Université de Montpellier
[8] Villa S, Stocco A, Blanc C and Nobili M 2020 Multistable interaction between a spherical Brownian particle and an air–water interface *Soft Matter* **16** 960–9
[9] Lançon P, Batrouni G, Lobry L and Ostrowsky N 2001 Drift without flux: Brownian walker with a space-dependent diffusion coefficient *EPL* **54** 28–34
[10] Volpe G, Helden L, Brettschneider T, Wehr J and Bechinger C 2010 Influence of noise on force measurements *Phys. Rev. Lett.* **104** 170602
[11] Magazzù A, Callegari A, Staforelli J P, Gambassi A, Dietrich S and Volpe G 2019 Controlling the dynamics of colloidal particles by critical Casimir forces *Soft Matter* **15** 2152–62
[12] Pesce G, McDaniel A, Hottovy S, Wehr J and Volpe G 2013 Stratonovich-to-Itô transition in noisy systems with multiplicative feedback *Nat. Commun.* **4** 2733
[13] Eliazar I and Klafter J 2005 On the nonlinear modeling of shot noise *PNAS* **102** 13779–82
[14] Batchelor G K 1976 Brownian diffusion of particles with hydrodynamic interaction *J. Fluid Mech.* **74** 1–29

IOP Publishing

Simulation of Complex Systems

Aykut Argun, Agnese Callegari and Giovanni Volpe

Chapter 8

The Vicsek model

Collective dynamics often emerge in living systems: fishes swim in schools, migrating birds fly in flocks, herds of cattle move together during seasonal migrations, bees fly in swarms. Even microorganisms, such as bacteria, sometimes swarm. And the motion of human crowds also exhibits collective dynamics. Swarm-like behaviors are instinctive in simpler species. In more complex species, they might result from conscious individual decisions. Often, a collective behavior entails some distinctive advantage both for the individuals and for the group. Moving in flocks, schools, and herds benefits individuals by providing more protection against predators (figure 8.1), or by reducing the energy cost of locomotion. Mutual coordination

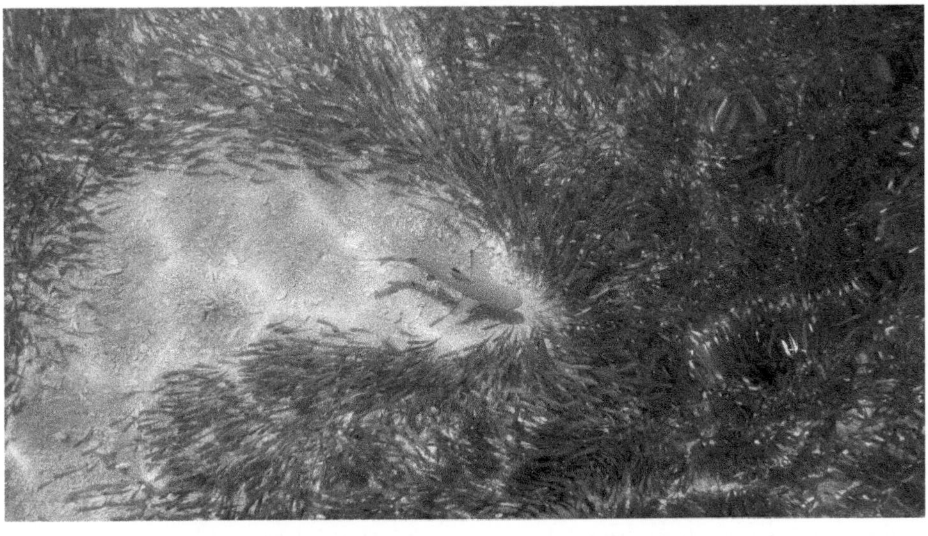

Figure 8.1. Living organisms self-organize. In Nature, we find several examples of self-organized living systems. Collective dynamics provides advantages for the species adopting it, for example, by increasing the chances that an individual will escape a predator, as in this photograph. *Source: Unsplash: Kris Mikael Krister.*

doi:10.1088/978-0-7503-3843-1ch8

and cooperation between individuals allow a group to have better chances of survival, for example, in harsh natural environments, preserving heat, energy, protecting offspring, deterring predators, and optimizing foraging.

Understanding the mechanisms that determine the emergence of collective motion and lead to swarm-like behavior is of the utmost importance in biology and ethology, but also has applications in technology, e.g., in the realization of robotic swarms for search-and-rescue operations or targeted collection and delivery, in biotechnology and medical engineering, to contain or eradicate the spread of pathogens and biological pests, and also in computer graphics, for realistic rendering of animal or crowd behavior. So, how do collective motion and swarm-like behaviors emerge in social organisms? How does the individual determine its motion on the basis of local sensory information about other individuals in its surroundings?

The *Vicsek model* is a minimalist model that accounts for swarm-like behavior [1]. The model itself is based on a very simple interaction rule: the individual, which moves at a constant speed, keeps on adjusting its direction to match the average direction of its neighbors. The Vicsek model has a second-order phase transition to a kinetic, swarm-like phase when it approaches a critical value of the noise parameter. By varying the noise level in the system, the density of the individuals, and the individual radius, the Vicsek model can be switched from a gas-like phase, in which the individuals move almost independently of each other, to a swarming phase, in which individuals self-organize in clusters.

In this chapter, we start from the standard definition of the Vicsek model and show how the different phases emerge as a function of the individual speed, interaction radius, noise level, and density of individuals. We then explore a few variations of the model, from slightly different rules for individuals to use to calculate their orientation, to the effect of delay in elaborating the sensory information from the surrounding individuals.

Example codes: Example Python scripts related to this chapter can be found at: https://github.com/softmatterlab/SOCS/tree/main/Chapter_08_Vicsek_Model. Readers are welcome to participate in the discussions related to this chapter on: https://github.com/softmatterlab/SOCS/discussions/17.

8.1 The standard Vicsek model

In the following, we will refer to each individual as a *particle* (they are sometimes also called *agents*). All N particles in the system are point-like and have the same *propulsion speed* v. The two-dimensional arena in which the particles move is a square with sides of length L centered at the origin of the axes, that has *periodic boundary conditions*: when a particle crosses over the right-hand boundary at $x = \frac{L}{2}$, it then re-enters from the left-hand side; and something analogous happens when the particle crosses any of the arena boundaries, as per the following rules:

$$\begin{cases} x \to x + L & \text{if} \quad x < -\frac{L}{2} \\ x \to x & \text{if} \ -\frac{L}{2} \leqslant x \leqslant \frac{L}{2} \\ x \to x - L & \text{if} \quad x > \frac{L}{2} \end{cases} \qquad \begin{cases} y \to y + L & \text{if} \quad y < -\frac{L}{2} \\ y \to y & \text{if} \ -\frac{L}{2} \leqslant y \leqslant \frac{L}{2}. \\ y \to y - L & \text{if} \quad y > \frac{L}{2} \end{cases} \tag{8.1}$$

Each particle interacts with all the other particles within a given distance R_f, called the *flocking radius*. Because of the periodic boundary conditions, each particle might also interact with particles located on the opposite side of the arena, as shown in figure 8.2. At time step $t_n = n\Delta t$, particle j has a position $r_{j,n} = (x_{j,n}, y_{j,n})$ and an orientation of $\theta_{j,n}$. This means that the particle is moving with a velocity vector given by $v_{j,n} = (v \cos \theta_{j,n}, v \sin \theta_{j,n})$. At each time step n, each particle collects information from all the particles in its flocking area and adjusts its orientation at the following time step $n + 1$ according to

$$\theta_{j,n+1} = \langle \theta_{k,n} \rangle + \eta W_{j,n} \Delta t, \tag{8.2}$$

where η is the *noise* and $W_{j,n}$ are random numbers with a uniform distribution on the interval $\left[-\frac{1}{2}, \frac{1}{2}\right]$ and the average is calculated for the particles within the flocking radius only, i.e., indexed by k with the condition $\left|\mathbf{r}_{k,n} - \mathbf{r}_{j,n}\right| < R_f$ (the particle j is itself included in the averaging). It is worth noting that the average of the angle is a *circular* average: $\left\langle \theta_{k,n} \right\rangle = \arctan\left[\left\langle \sin \theta_{k,n} \right\rangle / \left\langle \cos \theta_{k,n} \right\rangle\right]$. The velocity at time step t_n is then calculated by

$$\mathbf{v}_{j,n+1} = v\left(\cos \theta_{j,n}\, \hat{\mathbf{x}} + \sin \theta_{j,n}\, \hat{\mathbf{y}}\right) \tag{8.3}$$

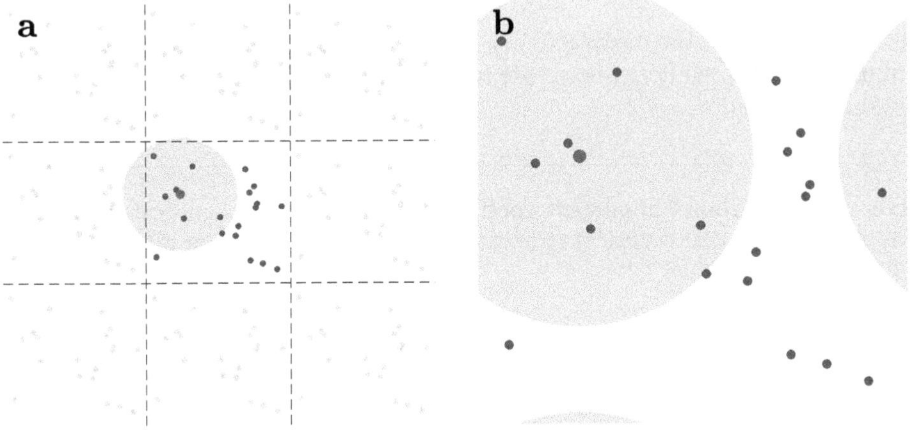

Figure 8.2. Two-dimensional arena with periodic boundary conditions. (a) A representation of the particles (full-color cyan dots) and their replicas (semi-transparent cyan dots) in the *unfolded* arena. For a given particle in the arena (larger red particle), the interaction area (or flocking area) is represented by the orange region. (b) The interaction area *folded* on the main arena. The large red particle (which is at the center of the interaction area) also interacts with a particle on the other side of the arena.

and it is used to calculate the position at t_{n+1}

$$\mathbf{r}_{j,n+1} = \mathbf{r}_{j,n} + \mathbf{v}_{j,n}\,\Delta t. \tag{8.4}$$

Therefore, only four parameters determine the behavior of the Vicsek model (excluding finite-size effects): the speed v, the flocking radius R_f, the particle density $\rho = \frac{N}{L^2}$, and the noise level η.

Exercise 8.1. Periodic boundary conditions. This first exercise is about implementing the correct functions for motion in an arena with periodic boundary conditions. Our system is made of N point-like particles in a square arena with side lengths of L. The interaction radius is R_f.

a. Generate a set of N random positions $\mathbf{r}_j$, with $j \in \{1, 2, \ldots N\}$ within the square arena. Displace all the particles by a small vector $\Delta\mathbf{r} = \Delta x\,\hat{\mathbf{x}} + \Delta y\,\hat{\mathbf{y}}$. Write a function that, given the new displaced position $\mathbf{r}'_j = \mathbf{r}_j + \Delta\mathbf{r}$, calculates the new correct position within the arena, implementing the periodic boundary conditions stated in equation (8.1).

b. Implement a function that, given the positions of N particles in a square arena with side lengths L, determines all the particles within a distance R_f from a given particle j with the position $\mathbf{r}_j$, taking into account the periodic boundary conditions, as represented by figure 8.2(b).

To monitor the global alignment of the system as time evolves, we define the *global alignment coefficient*:

$$\psi_n = \frac{1}{N}\left|\sum_{j=1}^{N}\frac{\mathbf{v}_{j,n}}{v}\right|, \tag{8.5}$$

which takes values in the interval [0, 1]. Values close to zero indicate the lack of any orientational order, while values close to one are an indication of a phase with high orientational order.

Exercise 8.2. Global alignment coefficient. Implement a function that calculates the global alignment coefficient (equation (8.5)) of a configuration of particles. Test your function on simple cases.

If we want to quantify the level of clustering of a configuration, we can either isolate the clusters one by one, counting the number of particles in each of them, or define a *clustering coefficient* in the following way. Given a configuration, we calculate its *Voronoi tessellation*. A Voronoi tessellation divides the arena into polygonal patches corresponding to each particle so that each patch includes all the points of the arena that are closer to the (only) particle included in the patch than to any other particle (most programming languages provide some built-in functions to calculate this). Given a particle i, let $A_{i,n}$ be the area of the Voronoi polygon for particle i at iteration n. We define the *global clustering coefficient* as:

$$c_n = \frac{\text{count}\left\{A_{i,n} < \pi R_{\text{f}}^2\right\}}{N}, \tag{8.6}$$

where only particles whose Voronoi cell is smaller than the flocking area are counted. Values of c close to zero indicate that there are few clusters. Values of c close to one indicate that most particles belong to a cluster, so there is a high level of clustering.

Exercise 8.3. Global clustering coefficient. Implement a function that calculates the global clustering coefficient (equation (8.6)) of a configuration of particles. Test your function on simple cases.

At this point, we have all the elements required to simulate the Vicsek model.

Exercise 8.4. Vicsek model at low noise and low density. In this exercise, take $L = 100$, $N = 100$, $v = 1$, $\Delta t = 1$, and $\eta = 0.01$. Implement your code and set the number of iterations to $S = 10^4$ steps. Compare your results with figure 8.3.

- **a.** Generate an initial particle configuration and save it in order to be able to start from exactly the same configuration for different flocking radii R_{f}. Plot the initial configuration.
- **b.** Perform the simulation for $R_{\text{f}} = 1$. Calculate and plot the global alignment and clustering coefficients as a function of the time step.

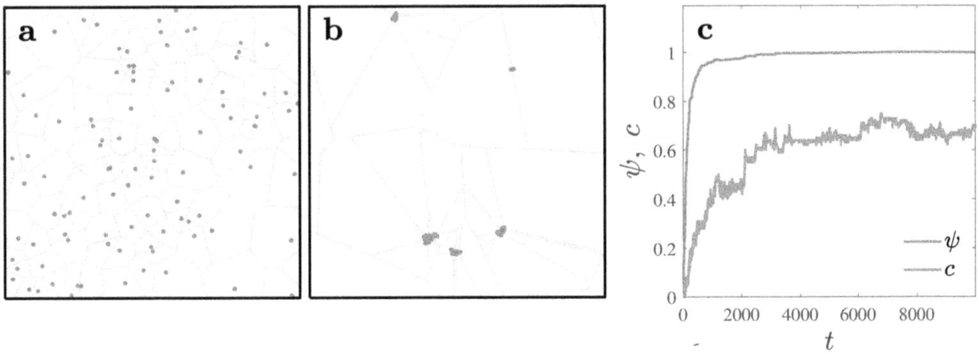

Figure 8.3. Vicsek model at low noise and low density. (a) Initial and (b) final configurations (after 10^4 steps, $N = 100$, $L = 100$, $\eta = 0.01$, $R_{\text{f}} = 1$, $v = 1$). The Voronoi tessellation is outlined by the cyan lines. (c) Global alignment coefficient ψ and global clustering coefficient c as a function of the time step. The global alignment coefficient shows that the particles are aligned after less than 1000 iterations. The clustering coefficient also steadily increases, but more slowly.

c. Plot the configurations after 10, 100, 500, and 1000 iterations, as well as the final configuration.

d. Repeat the simulation starting from the same initial configuration, but taking a larger detection radius (e.g., $R_f = 2, 5, 10, \ldots$).

When simulating the Vicsek model for low speeds, low densities, and low noise (exercise 8.4), you should have obtained the result that the particles organize into small clusters. Figure 8.3 represents the case for $R_f = 1$. If you repeat the simulation with a different value of R_f, for instance $R_f = 5$, you should notice that the particles, when clustering, tend to remain farther apart from each other. This is due to the alignment mechanism: if R_f is larger, particles start to align with particles further away.

Exercise 8.5. Increased noise. Repeat exercise 8.4 using a increased noise level, for example, $\eta = 0.1$. For ease of comparison, keep the other parameters the same. Run your simulation for $S = 10^4$ steps. Compare your results with figure 8.4.

With increased noise (exercise 8.5), you can see that the particles are slightly less aligned and slightly worse organized in clusters. If you increase the noise level even more, you should eventually find that the particles are no longer able to align and the clusters become more and more unstable until they cannot form at all.

Exercise 8.6. Higher density. Repeat exercise 8.4 using a higher density of particles, for example, $N = 1000$ (keep the other parameters the same). Run your simulation for $S = 10^4$ steps. Compare your results with figure 8.5.

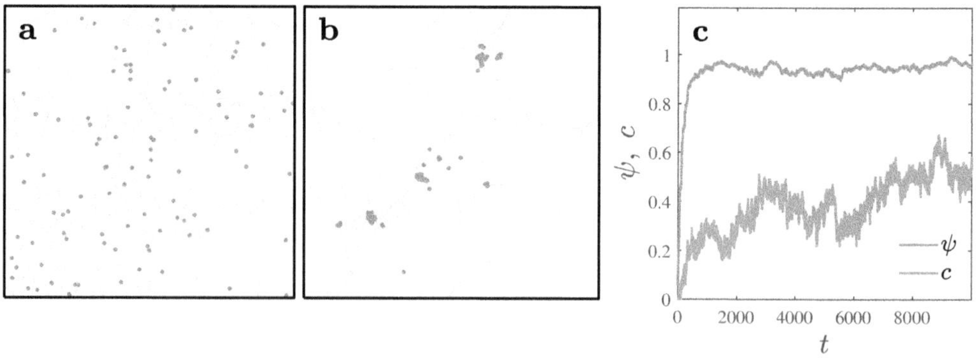

Figure 8.4. Vicsek model with increased noise. (a) Initial and (b) final configurations (after 10^4 steps) using the same parameters as in figure 8.3, but with increased noise ($\eta = 0.1$). (c) Global alignment coefficient ψ and global clustering coefficient c as functions of the time step. Compared to the case of less noise (figure 8.3), the global alignment coefficient ψ shows a slightly worse alignment, but it is the clustering coefficient that features the most evident changes, which demonstrate that the noise makes the clusters smaller and less stable.

Exercise 8.7. Higher density and increased noise. Now repeat exercise 8.4, increasing both the density (N = 1000) and the noise level ($\eta = 0.1$). Run your simulation and compare your results with figure 8.6.

When the density is increased (exercise 8.6), the particles rapidly show a phase with a high orientational order. With $R_f = 1$, the particles organize into several stable clusters. If you increase R_f, you will be able to obtain a single giant cluster including all the particles. Increasing the noise level (exercises 8.7) results in a slightly worse alignment and clustering (figure 8.6). If one increases the noise level

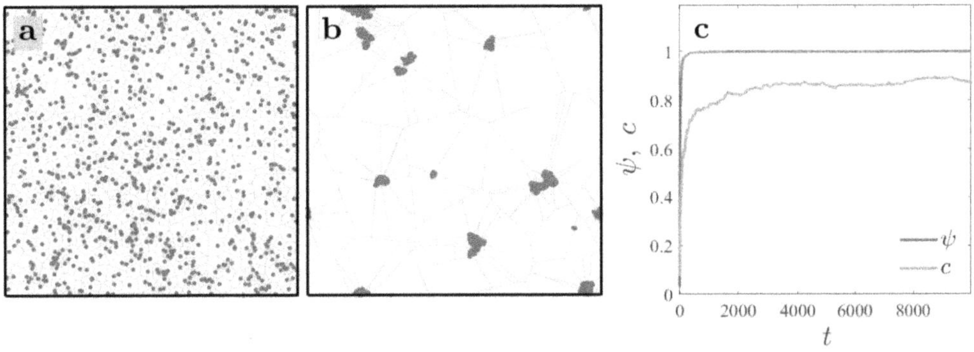

Figure 8.5. Vicsek model with higher density. (a) Initial and (b) final configurations (after 10^4 steps) with the same parameters as those used for figure 8.3, but a higher density (N = 1000). (c) Global alignment coefficient ψ and global clustering coefficient c as functions of the time step. Compared to the lower density case (figure 8.3), the global alignment coefficient ψ shows a faster and better alignment, and the clustering coefficient c fluctuates less and reaches a higher asymptotic value faster.

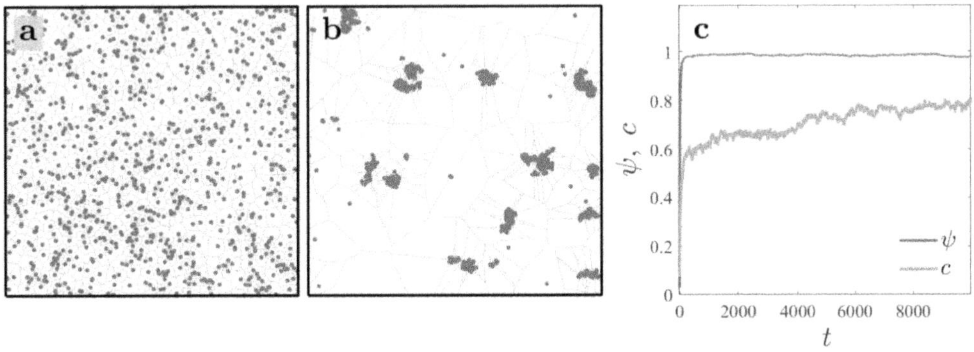

Figure 8.6. Vicsek model with increased noise and higher density. (a) Initial and (b) final configurations (after 10^4 steps) with the same parameters as those used for figure 8.3, but higher density (N = 1000) and increased noise ($\eta = 0.1$). (c) Global alignment coefficient ψ and global clustering coefficient c as functions of the time step. Compared to the case presented in figure 8.5 (the same density, but lower noise), the global alignment coefficient ψ does not change appreciably, but the clustering coefficient c fluctuates more and reaches a lower value.

even more for a high density, the particles move in an orientationally disordered phase, with only some correlation between the directions of nearby particles.

In the previous exercises, we have shown the data obtained from a single, long simulation. To gain insights into the statistical behavior of each of these cases (low density, high density, low noise, high noise), you should repeat the simulations with the same density and noise level settings, but different initial conditions, and take the average of the results.

8.2 The effect of delay

Sensory delays might occur when an individual applies rule (8.2) to re-orient itself. In such cases, rule (8.2) becomes

$$\theta_{j,n+1} = \langle \theta_{k,n-h} \rangle + \eta W_{j,n} \Delta t \tag{8.7}$$

where the integer h represents the delay. If $h = 0$, we have the standard Vicsek model.

Exercise 8.8. Delayed Vicsek model. Take $L = 1000$, $N = 100$, $\mathrm{v} = 3$, $\Delta t = 1$, $\eta = 0.4$, $R = 20$, so that the average number of particles within a detection area is not too high ($N_{\mathrm{f}} \approx 0.124$, because at higher densities, *density waves* might occur—check it out for yourself!). Moreover, the noise is strong enough to prevent the formation of stable clusters, but weak enough to allow some alignment of the orientations of the particles. With these parameters, a steady state is reached within 5000 time steps (usually much earlier).

a. Generate an initial configuration of positions and orientations. It might be convenient to have the same initial configuration for different values of the delay to ease their comparison.

b. Perform the simulation for $S = 10^4$ time steps for different values of the delay: Take, for example, $h = 0$ (no delay), 1, 2, 3, ⋯, 25. Check the evolution of the system in each simulation. Compare your results with figure 8.7.

c. For each simulation, calculate and plot the global alignment coefficient ψ and the global clustering coefficient c as functions of time, comparing the data for different delays h. Compare your result with figure 8.8.

d. Repeat these simulations for various initial conditions and statistically compare the results.

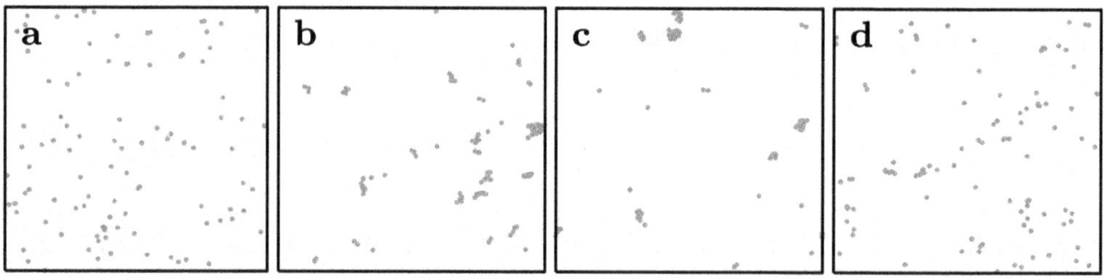

Figure 8.7. Delayed Vicsek model. Effect of delay on the Vicsek model ($N = 100$, $L = 1000$, $\eta = 0.4$, $\mathrm{v} = 3$). (a) Initial configuration. (b)–(d) Configurations after 1000 iterations for (b) $h = 0$ (no delay), (c) $h = 2$ (a short delay, which stabilizes the clustering), and (d) $h = 25$ (a long delay, which hinders the clustering).

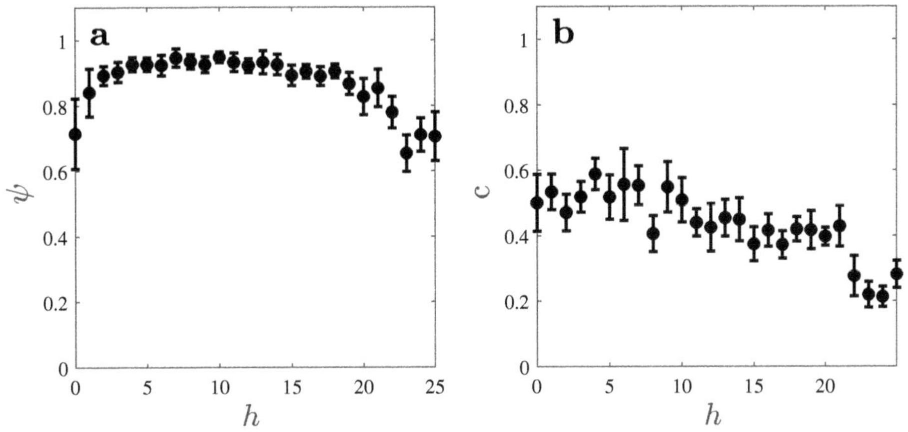

Figure 8.8. Effects of delay on alignment and clustering. (a) Global alignment coefficient ψ and (b) global clustering coefficient c as functions of the delay for single simulations (not averaged over different initial conditions). The data points are obtained by averaging ψ and c for the configurations after 5000 time steps. The parameters are the same as in figure 8.7.

A systematic study of the Vicsek model with delay is provided in reference [2], whose results show that the global alignment parameter is not drastically affected by the delay, but the clustering parameter is instead positively affected by the presence of a short delay, while for a longer delay, clustering is hindered.

One can also consider negative delays. We cannot directly use equation (8.7), because the expression $\theta_{k,n-h}$, with $h < 0$, refers to the angle at a time step that has not been calculated yet—it is a value in the future, after all. What is the meaning, then, of a negative delay? And how can we understand it? As we will see in chapter 10, a negative delay can be understood as a prediction: the individual, on the basis of the recent orientation history data, makes a prediction about the future value of the orientation of other individuals within its detection area, and uses this prediction to adjust its orientation. This is similar to when you want to move in a crowded place: you give a quick look around, you guess the immediate future trajectories of the other people by evaluating their current positions and velocities, and then you decide when and where to move, in order to avoid collisions. Thus, we can use a linear extrapolation of the orientation over the last s time steps to predict the orientation at a time $|h|\Delta t$ in the future. As we will see in the following exercise, the presence of a negative delay hinders cluster formation.

Exercise 8.9. Negative delays. Repeat exercise 8.8 for values of negative delay h between -1 and -15.

- **a.** Implement the procedure that linearly extrapolates the orientation $|h|$ steps into the future on the basis of the history of the last s orientations of each particle belonging to the detection radius. Take, for instance, $s = 5$
- **b.** Run the simulation for a negative delay $h = -1, -2, \ldots, -15$, and compare the results to those for the case of no delay ($h = 0$). In addition, treat the case of no

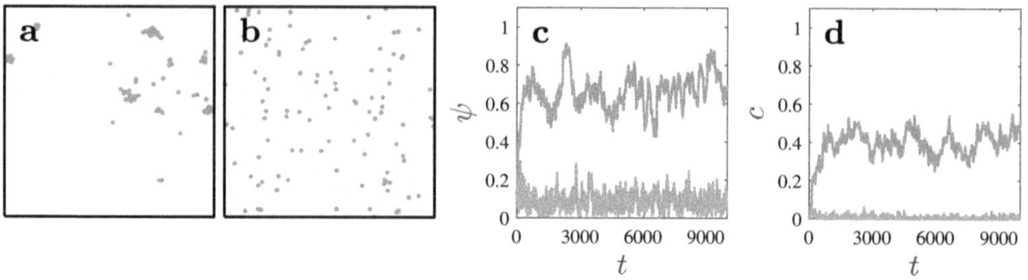

Figure 8.9. **The effect of negative delay.** Final configuration of the Vicsek model with (a) no delay ($h = 0$) and (b) negative delay $h = -2$ (i.e., using an extrapolation of the orientation of the particles two steps into the future). A comparison of the (c) global alignment coefficient ψ and (d) global clustering coefficient c for the two cases (the blue line represents no delay, and the red line represents a negative delay $h = -2$). In the presence of a negative delay, clustering is hindered.

delay using the extrapolation procedure. (Is this completely equivalent to the standard Vicsek model? Why?)

c. For each simulation, calculate the global alignment coefficient ψ and the global clustering coefficient c. Compare your results with figure 8.9.

8.3 Non-metric and non-reciprocal interactions

Flocks of birds have been extensively studied to understand whether their interactions follow the rules of the Vicsek model (equation (8.2)) and, if so, to estimate their interaction radius [3, 4]. Surprisingly, it has been found that birds usually take account of the orientations of their closest four to eight neighbors, to a large extent independently of their absolute distance from them, making the interaction *non-metric*. Furthermore, birds outside the field of view might not be seen and, therefore, taken into account, even if they are close by. In these cases, the interaction might turn out to be *non-reciprocal* (e.g., a bird might be influenced by another bird in front of it, but the bird in front is not influenced by the first bird). Therefore, it is interesting to study what happens to the predictions of the Vicsek model in the case of non-metric and non-reciprocal interaction rules. Here, we consider a specific non-metric and non-reciprocal interaction in which each individual considers only its k nearest neighbors to decide its orientation at the next step.

Exercise 8.10. Interaction with k nearest neighbors. Repeat exercise 8.4 by changing the interaction rule. In contrast to the standard Vicsek model, calculate the orientation by considering only the k nearest neighbors of a particle and the particle itself. Try different values of k (e.g., $k = 1, 4, 8$). Compare your results with figure 8.10. Verify that this interaction might be non-metric (i.e., the interaction does not only depend on the metric distance between particles) and non-reciprocal (i.e., if particle j

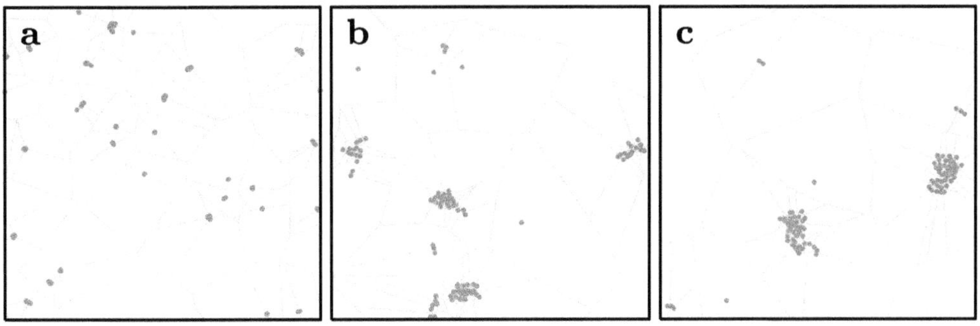

Figure 8.10. Vicsek model with k nearest neighbors. Configuration after 10^4 time steps for interactions with only (a) $k = 1$, (b) $k = 4$, and (c) $k = 8$ nearest neighbors ($N = 100$, $L = 100$, $\eta = 0.1$). Note how the size of the clusters increases with the value of k.

belongs to the k nearest neighbors of particle i, particle i is not necessarily included in the k nearest neighbors of particle j).

Exercise 8.11. Vision cones. Restrict the field of view of the particles in exercise 8.10 so that they can only see particles within a certain vision cone (e.g., only particles in front of them). Determine how this changes the overall behavior of the system.

8.4 Further reading

The Vicsek model was introduced in reference [1]. Its simplicity makes it the ideal benchmark for comparisons with experimental systems. For example, it was used in reference [5] for experimental investigations of the scaling laws in midget swarms and in reference [6] to model the information flow in finite active matter flocks as a function of noise. Reference [7] provides a review of the dynamics and finite-size scaling of the standard Vicsek model. Reference [8] covers the physics behind the Vicsek model, presenting the results of simulations together with the roles played by symmetries and conservation laws.

In the original article ([1]), it was claimed that the phase transition to a dynamical, collective ordered phase was a continuous (second-order) phase transition. This was later debated, due to the evidence produced by successive numerical simulations that indicated the phase transition could be discontinuous (first-order) [9]. However, further analysis eventually confirmed the continuous nature of the phase transition [10, 11].

Since delay and the imperfect transfer of information may play crucial roles in nature and technology, they must be taken into account to obtain a robust swarming mechanism. For the effect of delay in systems described by the Vicsek model, we refer the reader to reference [2]. For the application of the Vicsek model to the description of prey–predator systems, see reference [12].

The Vicsek model is not the only model that describes the swarming behavior of self-propelling particles. A more general approach is the Toner–Tu model [13], which includes the Vicsek model as a special case. However, it requires a more complex implementation. Other approaches simulate swarming behavior using data-driven approaches [14].

Examples of how experimental studies are conducted using real systems can be found in, for instance, references [3, 4, 15, 16]. Reference [17] explores theoretical models and empirical studies of collective behavior in animal groups, and also discusses its biological aspects. Another interesting reference for collective behavior is [18].

8.5 Problems

Problem 8.1. Vicsek with finite-radius particles. Simulate a system of self-propelled spherical colloids of radius R interacting according to equation (8.2). Consider the Brownian motion affecting the position of the particles and their finite size. Study the behavior of the system as a function of the strength of the aligning interactions and of the Brownian motion.

Problem 8.2. Vicsek with elongated particles. By analogy to problem 8.5.1, study the collective behavior of elongated Vicsek colloidal particles (with Brownian noise and steric repulsion) shaped in the form of prolate ellipsoids.

Problem 8.3. Vicsek in 3D. Simulate the Vicsek model in three spatial dimensions. [*Hint: Check references [19, 20].*]

Problem 8.4. Swarming in the presence of a predator. Introduce a predator into a prey swarm. Try to model the response of the prey in the proximity of the predator to obtain an effect like the one shown in figure 8.1. Rationalize and quantify how swarming improves the chances of an individual prey to survive an attack by a predator.

Problem 8.5. Continuous phase transition. Read reference [1] and replicate the results in the article. In particular, show the existence of a continuous phase transition in the Vicsek model by gradually decreasing the amount of noise η in simulation arenas of various sizes for a fixed density of particles ρ. You should obtain a transition from a disordered movement phase to a phase in which the particles move coherently.

Problem 8.6. Non-reciprocal interactions. Reference [21] studies the non-reciprocal interaction between two species of active colloid, A and B. The system consists of a suspension of spherical colloidal particles that, by means of a catalytic reaction on their surfaces, produce spherically-symmetric chemical gradients, which phoretically activate the other colloids in the solution. This mechanism leads to a non-reciprocal interaction, as the chemical produced by colloid A (B) induces phoretic repulsion (attraction) in colloid B (A). Translate this idea to a binary system of self-propelled particles with Vicsek-like interactions. Study the collective behavior of the system as a function of the relative fractions of the particles.

8.6 Challenges

Challenge 8.1. Sensory delay with non-reciprocal interactions. A swarm of small, disk-shaped robots with radius R move in a two-dimensional arena at a constant speed v. Each robot has a small sensor that can only perceive objects that are in front of them and located within an angle of $\|\alpha\| < \pi/2$. The processing of the image introduces a delay of t_d. Moreover, robots that are close may obstruct the view of other robots. The robots should try to avoid collisions with other robots. Try different alignment rules and study the collective behavior of such a system in an arena with reflecting boundary conditions.

Challenge 8.2. Delivery service with drones. You are responsible for the design of a delivery service for small items using autonomous drones. Each drone, when flying, has a target for the delivery and is aware of obstacles and other drones located in front of it within a certain range. Discuss the possible strategies for avoiding collisions and, at the same time, reaching the target within a determined amount of time. Also, discuss how the average delivery time scales with the number of active drones.

Challenge 8.3. Robots on uneven terrain. You are part of a team developing a search-and-rescue system based on autonomous robots. The terrain is uneven and some zones cause increased noise in the robot orientations. Study the behavior of a system of N identical robots. Determine which orientation rule is more appropriate to avoid collisions and to maximize the exploration of the area where the search-and-rescue operations must be performed.

References

[1] Vicsek T, Czirók A, Ben-Jacob E, Cohen I and Shochet O 1995 Novel type of phase transition in a system of self-driven particles *Phys. Rev. Lett.* **75** 1226–9

[2] Piwowarczyk R, Selin M, Ihle T and Volpe G 2019 Influence of sensorial delay on clustering and swarming *Phys. Rev.* E **100** 012607

[3] Ballerini M *et al* 2008 Interaction ruling animal collective behavior depends on topological rather than metric distance: Evidence from a field study *PNAS* **105** 1232–7

[4] Cavagna A *et al* 2010 From empirical data to inter-individual interactions: unveiling the rules of collective animal behavior *Math. Models Methods Appl. Sci.* **20** 1491–510

[5] Cavagna A *et al* 2017 Dynamic scaling in natural swarms *Nat. Phys.* **13** 914

[6] Brown J, Bossomaier T and Barnett L 2020 Information flow in finite flocks *Sci. Rep.* **10** 3837

[7] Baglietto G, Albano E V and Candia J 2012 Criticality and the onset of ordering in the standard Vicsek model *Interface Focus* **2** 708–14

[8] Ginelli F 2016 The physics of the Vicsek model *Eur. Phys. J. Spec. Top.* **225** 2099–117

[9] Grégoire G and Chaté H 2004 Onset of collective and cohesive motion *Phys. Rev. Lett.* **92** 025702

[10] Nagy M, Daruka I and Vicsek T 2007 New aspects of the continuous phase transition in the scalar noise model (SNM) of collective motion *Phys. A: Stat. Mech. Appl.* **373** 445–54

[11] Aldana M, Larralde H and Vázquéz B 2009 On the emergence of collective order in swarming systems: a recent debate *Int. J. Mod. Phys.* B **23** 3661–85

[12] Mohapatra S and Mahapatra P S 2019 Confined system analysis of a predator-prey minimalistic model *Sci. Rep.* **9** 11258

[13] Toner J and Tu Y 1995 Long-range order in a two-dimensional dynamical XY model: how birds fly together *Phys. Rev. Lett.* **75** 4326–9

[14] Ren J, Wang X, Jin X and Manocha D 2016 Simulating flying insects using dynamics and data-driven noise modeling to generate diverse collective behaviors *PLoS One* **11** 1–31

[15] Nagy M, Ákos Z, Biro D and Vicsek T 2010 Hierarchical group dynamics in pigeon flocks *Nature* **464** 890–3

[16] Ariel G and Ayali A 2015 Locust collective motion and its modeling *PLoS Comput. Biol.* **11** 1–25

[17] Giardina I 2008 Collective behavior in animal groups: theoretical models and empirical studies *HFSP J.* **2** 205–19

[18] Vicsek T and Zafeiris A 2012 Collective motion *Phys. Rep.* **517** 71–140

[19] Czirók A, Vicsek M and Vicsek T 1999 Collective motion of organisms in three dimensions *Phys. A: Stat. Mech. Appl.* **264** 299–304

[20] Gönci B, Nagy M and Vicsek T 2008 Phase transition in the scalar noise model of collective motion in three dimensions *Eur. Phys. J. Spec. Top.* **157** 53–9

[21] Soto R and Golestanian R 2015 Self-assembly of active colloidal molecules with dynamic function *Phys. Rev.* E **91** 052304

Aykut Argun, Agnese Callegari and Giovanni Volpe

Chapter 9

Living crystals

Many living organisms are motile: they self-propel by harvesting energy from their environment and converting it into directed motion. Motile living organisms can be found at all size scales, from tens of meters (e.g., whales) down to micrometers (e.g., cells, bacteria). Here, we focus on the microscopic scale, which is particularly important for biophysics and biomedicine, and learn to model the motion of microorganisms swimming in a fluid environment. Microorganisms can propel themselves using *active directed motion*, which is continuously disturbed by the presence of *thermal noise* (i.e., *Brownian motion*, see also chapter 5). Moreover, when several identical motile microorganisms come together, collective phenomena might emerge in which the microorganisms move in an organized, coordinated way, not unlike fishes in a shoal or birds in a flock (see also chapter 8).

Figure 9.1. The ancient Roman god Janus. Janus was the god of the duality of war and peace in Roman mythology. He was known for having two faces. One of the most common kinds of artificial microswimmer is a colloid with a gold coating on one side—a Janus particle. Image by *Quinn Dombrowski (CC BY-SA 4.0)*.

doi:10.1088/978-0-7503-3843-1ch9

Self-propulsion, emergent collective behaviors, and self-organization have also been reproduced by employing synthetic particles, known as *active particles*. These are microscopic colloidal particles capable of propelling themselves. If we consider a simple spherical colloidal particle, the key ingredient for self-propulsion is a breaking of symmetry around this particle. This can be achieved if the particle has different physical or chemical properties on its different sides. These particles are referred to as *Janus particles* after the two-faced Roman god of war and peace (figure 9.1). For example, a chemical reaction might take place on one side of the particle, generating some chemical products that act like a rocket propellant, causing the particle to move in a directed fashion.

Several mathematical models are used to describe the self-propelled motion of microscopic particles, which feature an interplay between directed motion and thermal noise. Two of the most common models are *active Brownian motion* and *run-and-tumble motion*. Active particles can form clusters known as *living crystals*. Living crystals are ordered, crystal-like structures that continuously rearrange themselves as if they were alive.

In this chapter, we learn how to simulate the equations governing active Brownian motion using a finite-difference numerical scheme (introduced in chapter 5). We then build on this simple model and consider a system with many active swimmers, which mutually interact only at short distances by *steric repulsion* and *phoretic forces*, showing how they form crystal-like, *metastable clusters* (i.e., living crystals). Finally, we introduce some longer-ranged *alignment interactions* between the active particles and show how these result in the formation of metastable clusters whose size and shape depend on the competition between the interaction strength and the noise level.

Example codes: Example Python scripts related to this chapter can be found at: https://github.com/softmatterlab/SOCS/tree/main/Chapter_09_Living_Crystals. Readers are welcome to participate in the discussions related to this chapter at: https://github.com/softmatterlab/SOCS/discussions/18.

9.1 Active Brownian motion

To model the behavior of biological and artificial microscopic self-propelling objects, or *microswimmers*, we introduce the concept of the *active Brownian particle*. Like Brownian particles (chapter 5), they are affected by thermal forces, whose action erratically alters their position and orientation. However, in contrast to Brownian particles, they feature directed motion.

Let us consider a single active Brownian particle characterized by a constant swimming speed v. This particle is subject to collisions with the molecules of the surrounding fluid, perturbing its trajectory and causing the particle to move randomly, generating *translational diffusion*. The orientation of this active particle also fluctuates due to these collisions, generating *rotational diffusion*. The resulting motion is described by the equations:

$$\begin{cases} \dfrac{\mathrm{d}x(t)}{\mathrm{d}t} &= v\cos\phi(t) + \sqrt{2D_\mathrm{T}}\,W_x \\ \dfrac{\mathrm{d}y(t)}{\mathrm{d}t} &= v\sin\phi(t) + \sqrt{2D_\mathrm{T}}\,W_y \\ \dfrac{\mathrm{d}\phi(t)}{\mathrm{d}t} &= \sqrt{2D_\mathrm{R}}\,W_\phi \end{cases} \tag{9.1}$$

where x and y are the particle's position, ϕ is its orientation (figure 9.2(a)), and D_T and D_R are the translational and rotational diffusion coefficients, respectively. W_x, W_y, and W_ϕ are independent random processes with a mean of zero and a variance of one. In a finite-difference numerical integration scheme, they are replaced by sequences of random numbers chosen from a Gaussian distribution with a mean of zero and an appropriate variance (as explained in detail in chapter 5). Some sample trajectories are shown in figure 9.2(b)–(e) for various values of v.

Exercise 9.1. Active Brownian particle. Write a program that generates and plots active particle trajectories (equation (9.1)) using a finite-difference method with a constant time step Δt (see chapter 5). Compare the results with figure 9.2.

- **a.** As a starting point, consider the following parameter values: $\Delta t = 0.01$ s, $D_\mathrm{T} = 2 \cdot 10^{-13}\,\mathrm{m^2\,s^{-1}}$, $D_\mathrm{R} = 0.5\,\mathrm{s^{-1}}$, $v = 3 \cdot 10^{-6}\,\mathrm{m\,s^{-1}}$. These are values in the ranges that are normally observed in experiments with artificial microscopic active particles or motile bacteria.
- **b.** Play with the parameters: what happens if you change the swimming speed v? Compare the plots for different values of v, as shown in figure 9.2(b)–(e).
- **c.** What happens if you change the value of the translational diffusion, D_T?
- **d.** What happens if you change the value of the rotational diffusion coefficient, D_R? Show that the motion of the particle becomes more directed (which makes it look qualitatively more active) as you decrease D_R.

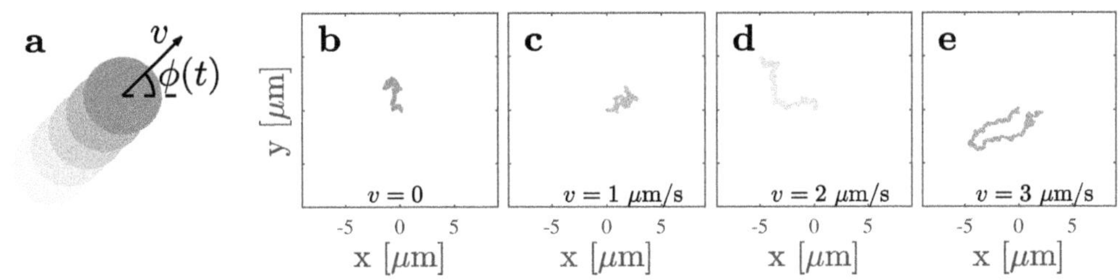

Figure 9.2. Active Brownian motion. (a) An active particle propels itself with a swimming velocity v that is subject to rotational diffusion (equation (9.1)). (b)–(e) Sample trajectories of active Brownian particles with swimming velocities of (b) $v = 0\,\mu\mathrm{m\,s^{-1}}$, (c) $v = 1\,\mu\mathrm{m\,s^{-1}}$, (d) $v = 2\,\mu\mathrm{m\,s^{-1}}$, and (e) $v = 3\,\mu\mathrm{m\,s^{-1}}$. The simulation parameters used are: $D_\mathrm{T} = 0.2\,\mu\mathrm{m^2\,s^{-1}}$, $D_\mathrm{R} = 0.5\,\mathrm{s^{-1}}$, 500 time steps, and $\Delta t = 10$ ms.

9.2 Mean square displacement

Let us start by considering the motion of a single active Brownian particle, as described by the stochastic differential equations (9.1). Even though each realization of this process generates a different stochastic trajectory, all these trajectories share some common characteristic features, which can be found by averaging. One of the most important quantities used to characterize the motion of an active particle is its *mean square displacement* (MSD):

$$\mathrm{MSD}(t) = \langle [x(t+\tau) - x(\tau)]^2 + [y(t+\tau) - y(\tau)]^2 \rangle, \tag{9.2}$$

where t is the time lag, $[x(t), y(t)]$ is a realization of the active motion resulting from equation (9.1), and the $\langle \dots \rangle$ is the *ensemble average*, i.e., the average over all the independent realizations of equation (9.1). The MSD can also be calculated via a *time average*, i.e., we use different starting points τ_i to obtain the average MSD from a single trajectory:

$$\mathrm{MSD}(t) = \frac{1}{n} \sum_{i=1}^{n} \langle [x(t+\tau_i) - x(\tau_i)]^2 + [y(t+\tau_i) - y(\tau_i)]^2 \rangle. \tag{9.3}$$

Since active Brownian motion is an *ergodic* process, the ensemble and time averages coincide; however, there are other processes for which this is not the case, some of which we have seen in chapter 6.

The general expression for the MSD of an active Brownian particle obeying the equations of motion (9.1) can be given in analytical form [1, 2]:

$$\mathrm{MSD}(t) = (4D_\mathrm{T} + 2v^2 t_\mathrm{r})t + 2v^2 t_\mathrm{r}^2 (e^{-t/t_\mathrm{r}} - 1), \tag{9.4}$$

where $t_\mathrm{r} = D_\mathrm{R}^{-1}$ is the characteristic timescale of the rotational diffusion, which represents the average time it takes the particle to change its direction of propagation. The functional dependence of the MSD on the time interval, then, allows us to discriminate, for example, particles that are actively moving from particles that are, instead, passive. In general, the MSD dependence has different characteristics for each different system.

You can see some sample trajectories in figure 9.2(b)–(e) and the related MSDs in figure 9.3. Note that active swimming results in superdiffusive behavior at certain timescales, i.e., the slope of log(MSD) vs log(t) is greater than one. As can be seen in figure 9.3, the MSD of an active Brownian particle is superdiffusive due to the ballistic motion (MSD $\propto v^2 t^2$) when $t \ll t_r$. In agreement with equation (9.4), when $t \gg t_r$, the particle undergoes enhanced diffusion (MSD $\propto t$) with an effective diffusion coefficient of $D_\mathrm{eff} = D_\mathrm{T} + v^2 t_r/2$.

Exercise 9.2. Mean square displacement of active Brownian particles. Using the code developed for exercise 9.1, generate some sample trajectories using different speeds and rotational diffusion constants.

a. Plot the trajectories and the related MSDs.

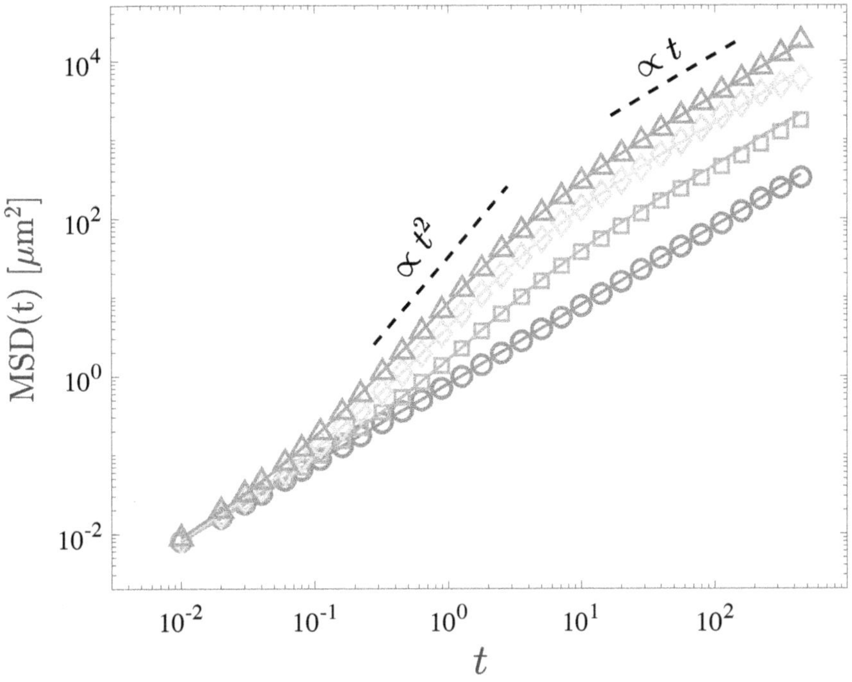

Figure 9.3. MSDs of active Brownian particles. MSDs (equation (9.2)) of active Brownian particles for $v = 0\,\mu\text{m s}^{-1}$ (blue circles, figure 9.2(b)), $v = 1\,\mu\text{m s}^{-1}$ (green squares, figure 9.2(c)), $v = 2\,\mu\text{m s}^{-1}$ (yellow diamonds, figure 9.2(d)), and $v = 3\,\mu\text{m s}^{-1}$ (orange triangles, figure 9.2(e)). The symbols represent the results of the simulations and the lines represent the theoretical results (equation (9.4)). The parameters are the same as those used for figure 9.2.

b. Compare the MSDs obtained using ensemble (equation (9.2)) and time averages (equation (9.3)). Verify that they coincide and conclude that this is a good indication that active Brownian motion is ergodic.

c. Can you think of a modified active Brownian motion for which the ensemble and time averages do not coincide? *[Hint: Consider processes that are not stationary.]*

9.3 Living crystals

We now consider a group of active, hard, spherical particles interacting only at short range via a phoretic attractive force. This model has been proposed to explain the experimental observations made in reference [3], in which a set of light-activated colloids, propelled by a phoretic force, formed metastable clusters that were called *living crystals.* The active building blocks of the clusters were microscopic polymer colloidal spheres with a protruding hematite cube, immersed in a basic solution containing hydrogen peroxide, which were activated via a blue light that triggered the decomposition of the hydrogen peroxide at the exposed hematite surface.

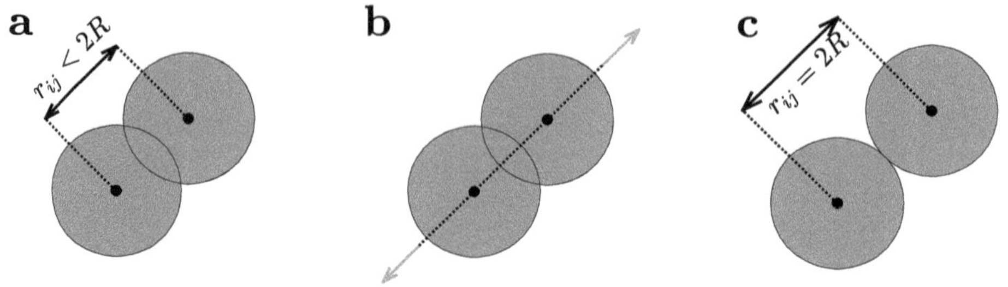

Figure 9.4. Volume exclusion for hard spheres. (a) When two spherical particles overlap during a simulation (i.e., the distance between their centers is smaller than their diameter $2R$), (b) we move them apart along the line joining their centers in opposite directions, (c) such that the final distance between them is $2R$.

Let us consider a system of many particles. To obtain a realistic simulation, one should take into account the fact that the particles should never overlap, i.e., occupy the same physical space. When two particles approach each other, collisions might happen. In the case of hard spherical colloids, this is well reproduced, in simulations, using the so-called *volume exclusion method*, or *hard-sphere correction* (figure 9.4): when an overlap between two particles is detected, then both particles are moved away from each other in the direction connecting their centers to a new configuration in which they touch each other, but are not superposed [4, 5]. Note that the common center of mass of the two particles remains in the same place before and after the particle rearrangement.

Exercise 9.3. Many-particle simulation. Write a program that generates and plots active particle trajectories, according to equation (9.1). Use a finite-difference method with a constant time step Δt for a system of N particles, and implement *volume exclusion.*

- **a.** First, write a function implementing the hard-sphere correction for a single pair of particles with a radius R, as shown in figure 9.4. Test your function: set the particles so that their centers are at a distance $d < 2R$, and check the new positions given by your function. *[Hint: The midpoint of the segment connecting the centers of the two particles should not have changed, and the two particles should touch each other, but not overlap, after the function has been applied.]*
- **b.** Using your newly-minted function for the hard-sphere correction, write some code to simulate the dynamics of $N = 100$ active particles. To begin with, use the following set of parameters, which are reasonable for a set of identical colloidal particles with $R = 1\ \mu\text{m}$ in water: $D_T = 0.1\ \mu\text{m}^2\text{s}^{-1}$, $D_R = 1\ \text{s}^{-1}$, $v = 3\ \mu\text{m s}^{-1}$, $\Delta t = 0.01$ s. Use periodic boundary conditions, choosing an appropriate box dimension to comfortably contain all the particles.
- **c.** How important is the choice of the time step Δt for the correct, physical behavior of the simulation? What if, for instance, you were to use $\Delta t = 10$ s? Is the choice of an appropriate Δt affected by the size of the particles? and by their speed?
- **d.** How is the optimal time step Δt affected by the concentration of particles?

Thanks to the volume exclusion method, we can simulate systems with more than one particle in a realistic way. We can proceed, then, to create some living crystals.

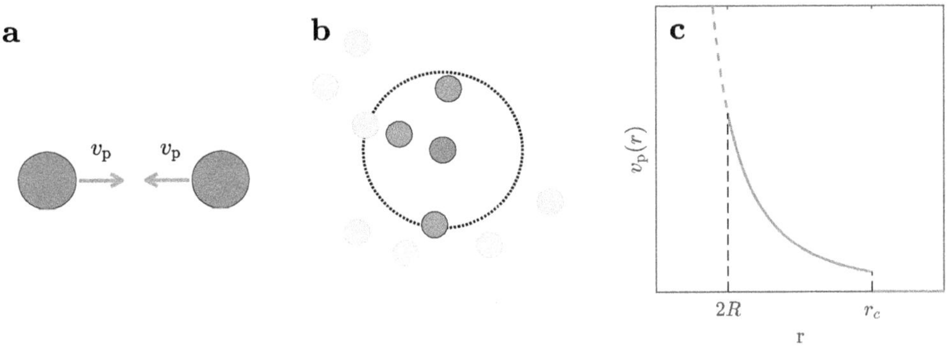

Figure 9.5. Phoretic interaction between active Brownian particles. (a) Two active Brownian particles (blue spheres) attract each other with a phoretic force that results in a drift velocity toward each other, denoted by v_p (orange arrows, equation (9.5)). (b) For faster execution of the simulation, phoretic interactions are only calculated for neighboring particles (orange spheres) that are closer than a cutoff radius (r_c, black dashed line). The particles that are outside this range (gray spheres) are ignored in the calculation of the phoretic forces acting on the blue particle. (c) The drift velocity due to phoretic attraction as a function of interparticle distance. Note that the minimum allowed interparticle distance is $2R$ due to volume exclusion and there is no interaction (i.e., $v_\mathrm{p} = 0$) if $r > r_\mathrm{c}$.

The interaction between the particles in reference [3] is phoretic in nature. It induces a force that is attractive, and gives the active particles a phoretic velocity $v_\mathrm{p}(r)$ that scales according to the inverse of the square of the distance between the particles:

$$v_\mathrm{p}(r) = v_0 \frac{R^2}{r^2}, \tag{9.5}$$

where v_0 is a constant velocity (i.e., a scaling factor for the phoretic interaction), R is the particle radius, and r is the interparticle distance. A schematic representation of the phoretic interaction is shown in figure 9.5(a). Since this force is significant only between nearby particles, a lot of computational time can be saved by assigning a cutoff radius (r_c, as shown in figure 9.5(b)). The strength of the resulting phoretic velocity as a function of the interparticle distance is shown in figure 9.5(c). This phoretic interaction is due to the advective flow generated by the decomposition of hydrogen peroxide on the exposed hematite surfaces of the light-activated colloids. The dependence on the inverse of the squared distance is the expected behavior for a phoretic attraction to a reaction source, and fits the behavior observed experimentally, as explained in reference [3].

Exercise 9.4. Active particles with phoretic interactions. Using the code implemented in exercise 9.3, write a program that generates and plots active particles interacting according to equation (9.5).

a. First write a function that implements the phoretic interaction between a single pair of particles. Your function should properly set the phoretic velocity of both particles as functions of the parameters v_0 and R which characterize the interaction. For the set of parameters given in exercise 9.3, start with $v_0 = 20\ \mu\mathrm{m\ s}^{-1}$.

b. Write a program to simulate the dynamics of many active particles interacting according to equation (9.5), using the hard-sphere correction to avoid overlap (what happens if you do not use this correction?). Choose a convenient time step Δt. Use a square box with sides $L = 100\ \mu\text{m}$ with $N \approx 50$ particles that have a radius $R = 1\ \mu\text{m}$ (periodic boundary conditions). Describe the long-term behavior of the system.

The strength of the interaction is determined by the magnitude of the reference speed v_0 at a given reference distance R. With increasing values of v_0, the interaction becomes stronger. In the following exercise, we will see that a stronger interaction causes the formation of larger, more stable clusters.

Exercise 9.5. Living crystals. Building on the code implemented in exercise 9.4, explore different values of v_0 and different concentrations of particles.

a. Use scaling factors for the phoretic interaction of $v_0 = 0\ \mu\text{m s}^{-1}$, $20\ \mu\text{m s}^{-1}$, and $50\ \mu\text{m s}^{-1}$. What is the behavior of the system after a long time as the interaction gets stronger, while the particle concentration and other parameters are kept constant? Compare your results to those presented in figure 9.6.

b. Vary the concentration of the particles (by changing either the number of particles or the dimension of the arena). Describe the ensuing behavior. Does a higher concentration of particles lead to larger clusters?

9.4 Aligning interactions

We now consider a system of active Brownian particles subject to *aligning interactions*. First, we focus on how these particles interact with each other, and then on how they form metastable clusters thanks to the interplay of their activity and alignment interactions [6].

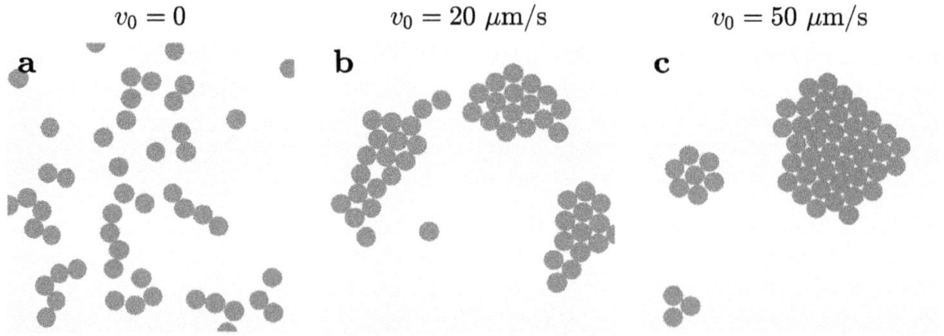

Figure 9.6. Living crystals. (a) When there is no phoretic interaction, active particles do not accumulate to create clusters. (b) When there is a weak phoretic interaction, the particles tend to come together and form small clusters. (c) When there is a strong phoretic interaction, the particles attract each other strongly and form large clusters.

The elementary particle–particle interaction is chosen such that it causes a torque on the interacting particles, and therefore changes their orientation. This interaction is considered to be additive; i.e., the total torque acting on a particle that is interacting with N particles is the sum of the elementary torques given by the individual pair interactions. Since the particles are in the overdamped regime, the effect of the total torque acting on each particle is an additional angular velocity obtained by using the angular equation from equations (9.1). The overall angular velocity of particle n is given by the following expression:

$$\omega_n = \omega_0 \, R^2 \sum_{i \neq n} \frac{\hat{v}_n \cdot \hat{r}_{ni}}{r_{ni}^2} \hat{v}_n \times \hat{r}_{ni} \cdot \hat{e}_z \qquad \text{for } r_{ni} < r_c, \tag{9.6}$$

where ω_0 represents the strength of interaction, R is the particle radius used as a distance scaling factor, $\hat{v}_n$ is the unit vector of the nth particle's instantaneous velocity, and $\hat{r}_{ni}$ is the unit vector from particle i to particle n. We only take alignment interactions into account for particles that are closer than a cutoff radius r_c (similar to the phoretic interactions, figure 9.5) which represents the region of interaction.

Figure 9.7 shows the dynamics of a system composed of two active particles for different values of the interaction strength ω_0 (for ease of comparison, $v = 10\ \mu\text{m s}^{-1}$ is kept constant, the initial configuration is the same, and there is no randomness, i.e., $D_T = 0$ and $D_R = 0$). For $\omega_0 = 0$ (figure 9.7(a)), the particles do not feel each other; their trajectories are straight. For $\omega_0 = 0.2\ \text{s}^{-1}$ (figure 9.7(b)), the particles barely feel each other: their trajectories bend a bit because of the interaction, but they do not end up forming a cluster. For $\omega_0 = 0.5\ \text{s}^{-1}$ (figure 9.7(c)), the effect of the interaction is more pronounced: the trajectories are bent much more, but, again, they end up not forming a cluster. For a very strong alignment interaction $\omega_0 = 1\ \text{s}^{-1}$ (figure 9.7(d)), the particles very quickly reorient to face each other, they head toward each other right after starting to interact, and they form a two-particle cluster.

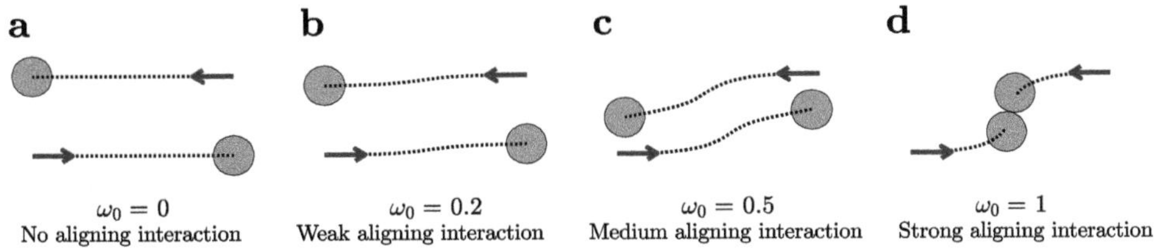

Figure 9.7. Alignment interactions. Two microswimmers ($R = 1\ \mu\text{m}$, $v = 10\ \mu\text{m s}^{-1}$) without diffusion ($D_T = 0$, $D_R = 0$) approach each other. (a) In the absence of alignment interactions ($\omega_0 = 0$ in equation (9.6)), the two particles pass each other without changing their orientation. (b) With weak alignment interactions ($\omega_0 = 0.2\ \text{s}^{-1}$), the particles turn slightly toward each other, but without a major impact on their overall trajectory. (c) With medium alignment interactions ($\omega_0 = 0.5\ \text{s}^{-1}$), the particles change their orientation toward each other significantly, but the distance between them is still too great for them to meet. (d) With strong alignment interactions ($\omega_0 = 1\ \text{s}^{-1}$), the particles quickly change their orientation toward each other as soon as they get close and move into contact with each other, eventually forming a two-particle cluster.

Exercise 9.6. Aligning interaction. Write some code to demonstrate the dynamics of two active particles interacting via the alignment interaction defined in equation (9.6). It might be useful to write a function that implements the alignment interaction defined in equation (9.6) between two particles only. Run the dynamics for different values of the interaction strength ω_0. Present your results similarly to those shown in figure 9.7.

If, instead of considering a system made of only two particles, we consider a system composed of a large number of particles, the aligning interactions can generate rich and complex collective behaviors, including the formation of metastable clusters whose size depends on the strength of the interaction and the noise level, as shown in figure 9.8.

Exercise 9.7. Metastable clusters. Generalize the code of exercise 9.6 to run with N particles (at least $N \approx 50$). You should also be able to plot the configuration of the particles at a chosen time.

a. Using the values $\omega_0 = 0.5\ \text{s}^{-1}$, $v = 10\ \mu\text{m s}^{-1}$, $r_c = 10R$, and $D_R = 0.1\ \text{s}^{-1}$, identify individual clusters and plot some sample screenshots of the simulation for different cluster sizes. Make a video that displays the time evolution of this system.

b. Play with the interaction strength ω_0, changing its value to $\omega_0 = 0.1\ \text{s}^{-1}$ and $\omega_0 = 1\ \text{s}^{-1}$. What do you observe in the collective dynamics of the particles?

c. Play with the rotational diffusion while keeping the interaction strength ω_0 constant. For example, try $D_R = 0.01\ \text{s}^{-1}$ and $D_R = 1\ \text{s}^{-1}$. What do you observe in the collective dynamics of the particles? Compare your results to figure 9.8.

d. Express how the strength of the interaction ω_0 and the rotational diffusion constant D_R affect the collective dynamics of the active particles. Provide supporting numerical evidence for your statements. *[Hint: For more details, see reference [6].]*

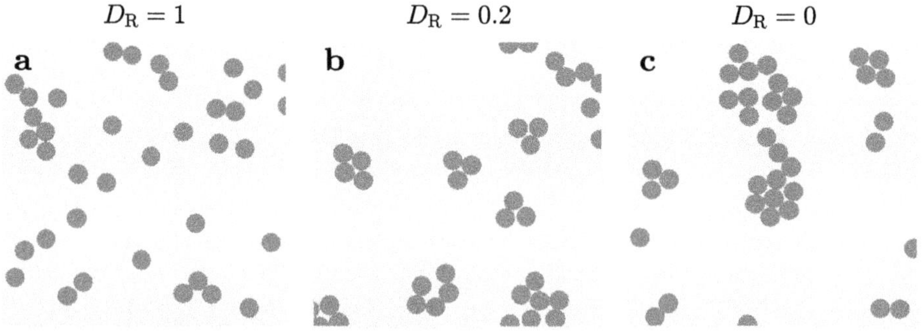

Figure 9.8. Metastable clusters. (a) When the rotational diffusion coefficient is large ($D_R = 1$), the clustering behavior is not observed. (b) When the rotational diffusion is lower ($D_R = 0.2$), the particles form small clusters. (c) When there is no rotational diffusion ($D_R = 0$), larger clusters appear. In all cases, the active Brownian particles have strong alignment interactions ($\omega_0 = 50\ \text{s}^{-1}$).

A natural further step is to introduce passive particles into this system. In the next exercise, we will see that it is possible to define an active–passive alignment interaction such that, in the case of active particles swimming in an environment of densely packed passive particles, metastable channels are formed by the active agents [6].

Exercise 9.8. Metastable channels. Simulate a mixture of active and passive particles in a densely packed environment. Active particles should have alignment interactions with each other. In contrast to active–active interactions, interactions between an active and a passive particle should have the opposite sign, so that a swimming particle will tend to 'turn away' when approaching a passive one. Create this model and find parameters that exhibit metastable channels such as those shown in figure 9.9; present your phenomenon using an animation.

9.5 Further reading

Reference [2] provides a comprehensive review of activity at the microscopic level in its experimental and theoretical aspects, including the effects of a complex or crowded environment. The physics of active motion is explained using the theoretical framework of stochastic thermodynamics and non-equilibrium thermodynamics, which is comprehensively reviewed in reference [7]. The mathematical models used to describe different kinds of self-propelling swimmer, disregarding microscopic details, can be found in references [8, 9], where the implications for collective behaviors are also discussed.

Most of the exercises in this chapter are inspired by references [3, 6, 10]. You may refer to these papers for further inspiration and details of the simulations.

For information about the simulation of active Brownian particles with chirality, reference [11] is a great resource, where you can also find novel methods for sorting microswimmers based on their chirality.

Specific investigations of the behavior of artificial microswimmers, often inspired by biological swimmers, are given in references [1, 12–14].

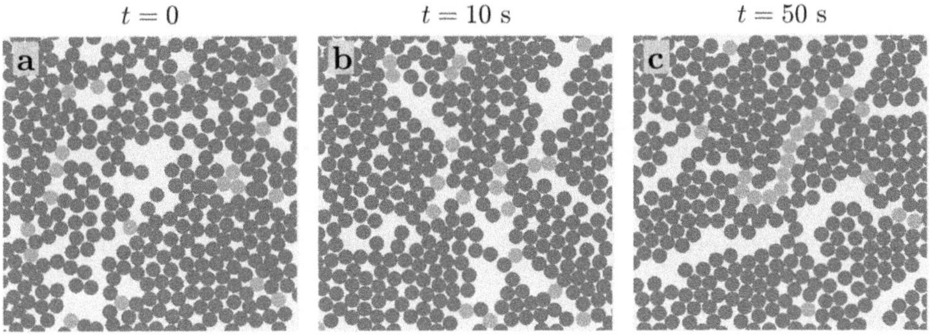

Figure 9.9. Metastable channels. A simulation of 20 active particles ($v = 40\ \mu\text{m s}^{-1}$, orange disks) and 350 passive particles ($v = 0\ \mu\text{m s}^{-1}$, blue disks) of radius $R = 4\ \mu\text{m}$ placed in an ($80\ \mu\text{m} \times 80\ \mu\text{m}$) environment with an alignment interaction strength of $\omega_0 = 10\ \text{s}^{-1}$. Screenshots of the simulation at (a) $t = 0$ s, (b) 10 s, and (c) 50 s. The active particles form and follow channels while moving through the dense environment of passive particles.

9.6 Problems

Problem 9.1. Mean square displacement of the particle orientation. Can you think of a definition for the *mean square angular displacement* (MSAD)? Try to think by analogy, considering the definition of the MSD given in equation (9.2).

a. Write some code to calculate the MSAD from the results of an active Brownian simulation. Repeat the simulation and the calculation of the MSAD for different values of the rotational diffusion constant D_R. Plot the MSAD obtained in a graph similar to figure 9.3.

b. Repeat the same calculations with time averaging (equation (9.3)). Show that the rotational diffusion of an active Brownian particle is ergodic.

Problem 9.2. Active Brownian particle in a circular well. Consider an active Brownian particle that is confined to a circular well with a radius of L.

a. Show that the position distribution of the particle changes from a flat distribution to an edge-peaked distribution as you increase the swimming velocity v. Can you see why this is a signature that the active particle is no longer in thermodynamic equilibrium?

b. Calculate the MSD of the particle as a function of time. Show that, unlike a free active Brownian particle, the MSD of a confined particle saturates.

Problem 9.3. Dancing active particles. Consider two active Brownian particles with radius R that are connected by a rigid rod of length L. *[Hint: Simulate the motion of both particles according to equation (9.1)), but impose the condition that they must always be at the same distance at the end of each time step.]*

a. Calculate the orientation of the rod ϕ at each step and plot this as a function of time.

b. Calculate the rotational diffusion coefficient of the rod connecting the particles and plot this as a function of R/L.

c. Calculate the effective swimming velocity and the rotational diffusion coefficient of these particles.

Problem 9.4. Active Brownian motion in a harmonic trap. Simulate the motion of an active Brownian particle in a harmonic trap. *[Hint: Generalize equation (5.9).]*

a. Observe the distribution of the particle inside the trap as a function of the swimming velocity v.

b. Assume that the size of the harmonic potential has finite radius. Calculate the swimming velocity that would allow the particle to escape the potential.

Problem 9.5. Negative aligning interactions. Simulate a system of active particles with negative alignment interactions ($\omega_0 < 0$).

a. Show that the particles avoid each other.

b. Introduce a positive phoretic attraction along with the negative alignment interaction. Explore the competition between these two interactions by playing with the parameter range.

Problem 9.6. Chiral active motion. Consider an active Brownian particle with chirality. This means that the particle will have a constant rotational velocity that changes its orientation in addition to the rotational diffusion. In mathematical terms, the angular equation in equation (9.1) will change to

$$\frac{d\phi(t)}{dt} = \sqrt{2D_R}\ W_\phi + \omega,$$

where ω is the angular velocity that is associated with the chiral motion.

a. Calculate the MSD of a chiral active particle as a function of time and observe its behavior for different values of ω.

b. Show that particles with different chirality signs can be sorted using 'chiral flowers' of elliptical obstacles, as described in reference [11].

9.7 Challenges

Challenge 9.1. Active particle sorting by bracket-shaped obstacles. Consider a box with active particles that are separated by bracket-like obstacles. Show that this can be used as a filter for active particles, as active particles (as opposed to passive Brownian particles) will eventually gather at one side of the box. For details, see figure 6 of reference [10].

Challenge 9.2. Rectification of active Brownian particles in a ratchet. Design a ratchet-like channel (a stack of cut triangle areas that are connected; for details see figure 5 of reference [10]). Show that a standard Brownian particle released in such a channel remains, on average, at the initial position, while active particles, on average, acquire a directed motion.

References

[1] Howse J R *et al* 2007 Self-motile colloidal particles: from directed propulsion to random walk *Phys. Rev. Lett.* **99** 048102

[2] Bechinger C *et al* 2016 Active particles in complex and crowded environments *Rev. Mod. Phys.* **88** 045006

[3] Palacci J, Sacanna S, Steinberg A P, Pine D J and Chaikin P M 2013 Living crystals of light-activated colloidal surfers *Science* **339** 936–40

[4] Schaertl W and Sillescu H 1994 Brownian dynamics of polydisperse colloidal hard spheres: equilibrium structures and random close packings *J. Stat. Phys.* **77** 1007–25

[5] Callegari A and Volpe G 2019 Numerical simulations of active Brownian particles *Flowing Matter, Soft and Biological Matter* ed F Toschi and M Sega (Springer: Cham) 211 p

[6] Nilsson S and Volpe G 2017 Metastable clusters and channels formed by active particles with aligning interactions *New J. Phys.* **19** 115008

[7] Seifert U 2012 Stochastic thermodynamics, fluctuation theorems and molecular machines *Rep. Prog. Phys.* **75** 126001

[8] Ramaswamy S 2010 The mechanics and statistics of active matter *Annu. Rev. Condens. Matter Phys.* **1** 323–45

[9] Romanczuk P, Bär M, Ebeling W, Lindner B and Schimansky-Geier L 2012 Active Brownian particles *Eur. Phys. J. Spec. Top.* **202** 1–162

[10] Volpe G, Gigan S and Volpe G 2014 Simulation of the active Brownian motion of a microswimmer *Am. J. Phys.* **82** 659–64

[11] Mijalkov M and Volpe G 2013 Sorting of chiral microswimmers *Soft Matter* **9** 6376–81
[12] Jiang H-R, Yoshinaga N and Sano M 2010 Active motion of a Janus particle by self-thermophoresis in a defocused laser beam *Phys. Rev. Lett.* **105** 268302
[13] Volpe G, Buttinoni I, Vogt D, Kümmerer H-J and Bechinger C 2011 Microswimmers in patterned environments *Soft Matter* **7** 8810–15
[14] Buttinoni I, Volpe C, Kümmel F, Volpe G and Bechinger C 2012 Active Brownian motion tunable by light *J. Phys. Condens. Matter* **24** 284129

Aykut Argun, Agnese Callegari and Giovanni Volpe

Chapter 10

Sensory delay

Various natural and artificial systems consist of multiple autonomous agents that communicate with each other and work collaboratively. Examples include rescue robots [1], bacterial colonies [2, 3], warehouse trucks [4], and human crowds [5]. The interaction between these agents often creates complex behaviors. An important parameter of such systems is the *sensory delay* between the time when an agent receives a signal and when it reacts to it. Depending on the system, the reason for this delay can be optical, chemical, or biological. It is also possible that the delay is added intentionally to engineer a desired system behavior.

Autonomous robots represent typical examples of autonomous agents (figure 10.1). They are widely employed in industry and operate at different scales. A robot

Figure 10.1. Swarming of robots. Each robot emits light and adjusts its speed depending on the total light intensity it measures. By controlling the sensory delay, it is possible to engineer the behavior of a group of robots, getting them to cluster together or to disperse away from each other—a feature that can be interesting for many applications, e.g., in delivery and search-and-rescue operations. *Picture by Mite Mijalkov.*

doi:10.1088/978-0-7503-3843-1ch10

fundamentally has two sets of elements: *sensors* and *motors*. While sensors provide the robots with information about their surroundings, motors enable them to move or do work in their environment. The introduction of a sensory delay can alter the behavior of the robots, at both the individual and group levels [6].

In this chapter, we simulate the behavior of robots with sensory delay. First, we analyze the response of a single robot to a light field as a function of sensory delay. We then simulate a group of robots and show that the clustering behavior of these robots depends strongly on their sensory delay. Specifically, we show that the robots' sensory delay can be tuned to make them cluster together or explore their environment by dispersing away from each other.

> **Example codes:** Example Python scripts related to this chapter can be found at: https://github.com/softmatterlab/SOCS/tree/main/Chapter_10_Sensory_Delay. Readers are welcome to participate in the discussions related to this chapter at: https://github.com/softmatterlab/SOCS/discussions/19.

10.1 A light-sensitive robot

First, we consider a single robot that moves in a two-dimensional plane while subject to rotational diffusion. The equations of motion for this autonomous robot can be written as follows:

$$\begin{cases} \dfrac{\mathrm{d}x(t)}{\mathrm{d}t} = v \cos \phi(t) \\ \dfrac{\mathrm{d}y(t)}{\mathrm{d}t} = v \sin \phi(t) \\ \dfrac{\mathrm{d}\phi(t)}{\mathrm{d}t} = \sqrt{\dfrac{2}{\tau}} w_\phi \end{cases} \tag{10.1}$$

where (x, y) are the position of the robot, ϕ is its orientation, v is its speed, τ represents its rotational diffusion timescale, and w_ϕ is a white noise. A schematic representation is shown in figure 10.2(a). The robot adjusts its speed v according to the light intensity that it measures: it changes its speed from V_0 to V_{inf} using an exponential decay function:

$$v = V_{\mathrm{inf}} + (V_0 - V_{\mathrm{inf}})e^{-I}, \tag{10.2}$$

where I is the measured light intensity. The robot moves at its maximum speed (V_0) if the measured light intensity is zero and exponentially slows down to the minimum speed (V_{inf}) as the measured light intensity gets higher, as shown in figure 10.2(b).

Although we analyze the case of an autonomous robot here, a variety of real-world agents perform similar motions. Examples include microswimmers, motile bacteria, and animals. All these agents adjust their speed according to the sensory inputs they receive from their environment. After all, at a constant speed, the robot

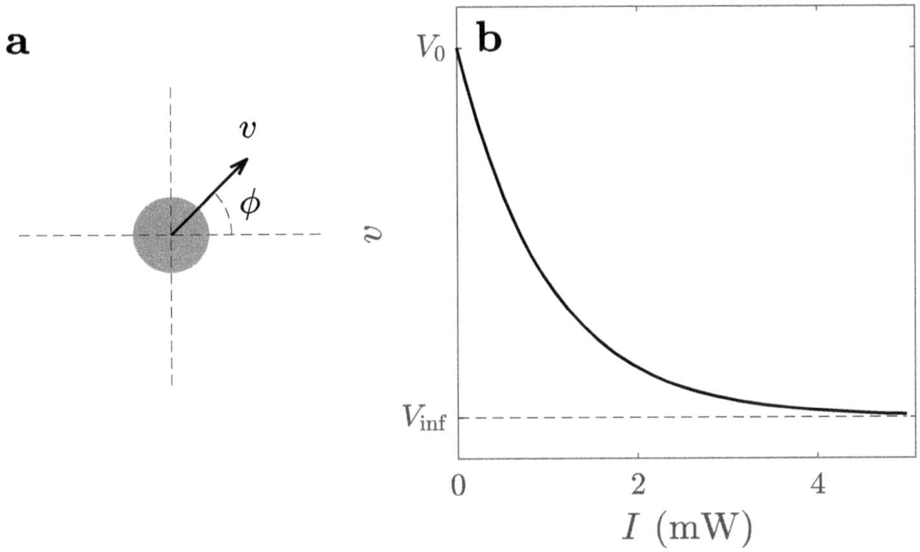

Figure 10.2. Working principle of a light-sensitive robot. (a) The motion of the robot is described by equation (10.1). It moves with a speed v that is a function of the optical intensity it measures, as shown in (b), in a direction ϕ that randomly changes over time. (b) The robot's speed as a function of the measured light intensity (equation (10.2)).

would, in practical terms, perform an active Brownian motion (see the equations for active Brownian motion in chapter 9).

Exercise 10.1. Simulation of a light-sensitive robot. Simulate the motion of a robot (equation (10.1)) with light-sensitive speed (equation (10.2)).

a. Simulate the motion of the robot in the absence of light (i.e., $I \equiv 0$ and $v \equiv V_0$). Show that its motion is standard active Brownian motion with a persistence length $L = V_0\tau$.

b. Simulate the motion of a robot in a periodic light pattern given by

$$I(x) = \left[\sin\left(2\pi\frac{x}{\Lambda}\right)\right]^2.$$

Study the robot's motion as a function of Λ/L, for instance choosing values of Λ between $L/10$ and $10\ L$. Where does the robot spends most of its time? In the bright or in the dark fringes?

c. Now consider a time-varying light pattern given by

$$I\left(x,\ t\right) = \left[\sin\left(2\pi\frac{x - ct}{\Lambda}\right)\right]^2.$$

Show that this light pattern induces an average motion of the robot in the x direction. Study the robot's motion as a function of the parameter c (it can be useful to consider the relation between the characteristic length scales $c\tau$, Λ, and L). What is the value of c that maximizes the robot's motion? Does the average robot motion change direction as a function of the parameters?

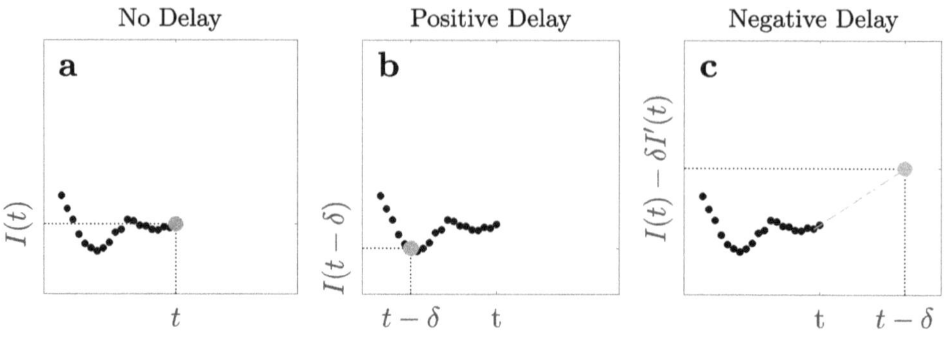

Figure 10.3. Null, positive, and negative sensory delays. (a) When the robot reacts to the measured light intensity (black dots) with *no delay* ($\delta = 0$), it uses the currently measured value $I(t)$ (blue dot) to adjust its speed. (b) When the robot reacts to the measured light intensity with *positive delay* ($\delta > 0$), it uses the past value measured an amount of time δ previously $I(t-\delta)$ (orange dot). (c) When the robot reacts with *negative delay* ($\delta < 0$), it uses the extrapolated future value of the light intensity $I(t-\delta) \approx I(t) - \delta I'(t)$ (green dot).

10.2 Single robot with a sensory delay

So far, we considered the case in which the measured light intensity affects the robot's speed immediately. In this section, we consider that the propulsion speed of the robot v is a function of the measured light intensity with a delay:

$$v = v(I(t - \delta)), \tag{10.3}$$

where $I(t-\delta)$ refers to the light intensity that the robot measured an amount of time δ previously. We will observe that the robot's qualitative response to a light field can be tuned just by changing this delay parameter.

If we have *no sensory delay* ($\delta = 0$), the robot uses the current value of the light intensity it measures ($I(t)$) instantly, as shown in figure 10.3(a). If we have a *positive delay* ($\delta > 0$), the robot reacts to the value of the past intensity that it measured an amount of time δ previously, as shown in figure 10.3(b). Although less intuitive, it is also possible to have a *negative delay* ($\delta < 0$): The robot can use the current intensity and its trend (or time derivative) in order to approximate a future state of the light intensity with which to adjust its speed, i.e., $I(t-\delta) = I(t) - \delta I'(t)$ for $\delta < 0$, as shown in figure 10.3(c). This can be interpreted as the action of the robot in response to the predicted future state of the system. Please note that, when linearly extrapolating the value of the intensity, the extrapolation should not lead to negative light intensity measured.

In this section, we will focus on how the delay parameter δ can be engineered to control the behavior of an autonomous robot. Specifically, a positive sensory delay ($\delta > 0$) makes the robot spend more time in brighter (i.e., lower-speed) regions and a negative sensory delay ($\delta < 0$) makes the robot escape from brighter (i.e., lower-speed) regions.

Exercise 10.2. Robot in a Gaussian light intensity zone. Simulate the motion of a light-sensitive robot (equation (10.1)) in a Gaussian light intensity profile (as shown in figure 10.4(a))

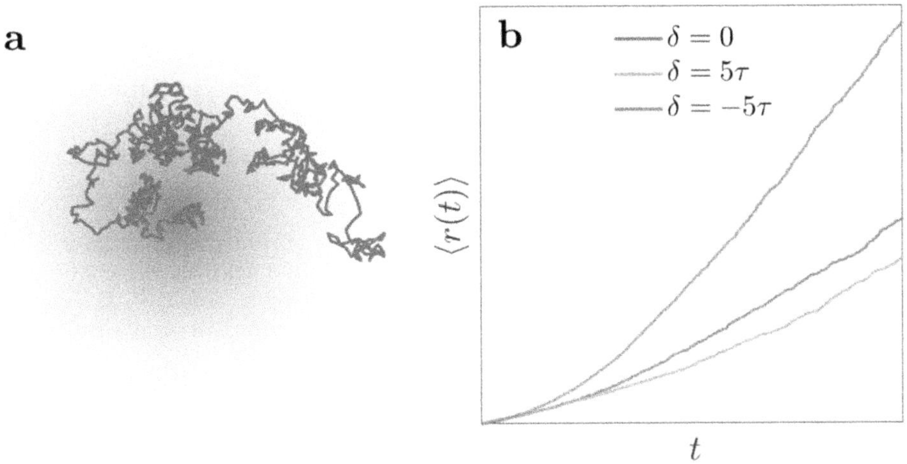

Figure 10.4. Robot in a Gaussian light field with a sensory delay. (a) An autonomous robot that moves according to equation (10.1) is placed in a Gaussian light field starting from the center. The statistical properties of the robot's motion can be tuned by controlling the delay, δ. A sample trajectory with no sensory delay ($\delta = 0$) is shown (blue line). (b) The average distance of the robot from the center as a function of time for no delay (blue line, $\delta = 0$), for positive delay (red line, $\delta = 5\tau$), and for negative delay (green line, $\delta = -5\tau$). The robot with a positive sensory delay tends to stay closer to the high-intensity region, whereas the robot with a negative sensory delay escapes the high-intensity region quickly. The parameters used for this simulation are $I_0 = 1$ W, $r_0 = 1$ m, $V_0 = 1$ m s^{-1}, $V_\infty = 0.1$ m s^{-1}.

$$I\left(x, y\right) = I_0 e^{-(x^2+y^2)/r_0^2},$$

where $I_0 = 1$ W and $r_0 = 1$ m are constants. Use the exponential decay function in equation (10.2) (figure 10.2(b)) with a delay for the particle's response to the light field, i.e.,

$$V(I, \delta) = V_{\text{inf}} + (V_0 - V_{\text{inf}})e^{-I(t-\delta)},$$

where $I(t - \delta)$ is the light intensity measured by the robot with a sensory delay of δ.

a. Simulate the motion of a robot with no sensory delay ($\delta = 0$). Start by placing the robot in the middle of the Gaussian light field ($x(0) = 0$, $y(0) = 0$). Plot some example trajectories, as shown in figure 10.4(a).

b. Play with the parameter δ and visualize the differences in the robot's behavior. Observe that the robot is more likely to stay longer in the brighter region for positive delay values and escapes quicker for negative delays.

c. Measure and plot the average radial distance from the center $\langle r(t) \rangle$ with $r(t) = \sqrt{x(t)^2 + y(t)^2}$ for different delays ($\delta = 0$, $\delta = 5\tau$ and $\delta = +5\tau$). Show the quantitative difference, as illustrated in figure 10.4(b).

So far, we have studied the behavior of the robot in an unbounded environment. However, boundaries are often present. In the next exercise, we are going to examine the positional distributions of the robot as a function of sensory delay in a bounded environment.

Exercise 10.3. Robot in a circular well. Repeat the simulation in exercise 10.2 ($I_0 = 1$ W, $r_0 = 1$ m, $V_0 = 1$ m s^{-1}, $V_\infty = 0.1$ m s^{-1}) but with a solid circular boundary of radius $R = 2$, as shown in figure 10.5(a). *[Hint: The boundary can be implemented by, at each time step, checking whether the robot is outside the well (i.e.,* $x(t)^2 + y(t)^2 > R^2$*) and, if so, reflecting its position back into the well (i.e.,* $x(t) \to x(t)R/\sqrt{x(t)^2 + y(t)^2}$, $y(t) \to y(t)R/\sqrt{x(t)^2 + y(t)^2}$*).]*

a. Calculate the spatial probability distribution of the robot as a function of the delay. Show that the robot tends to spend more time near the center (borders) when the delay is positive (negative), as shown in figures 10.5(b)–(d). Can you find a value of the delay for which the probability distribution is uniform? *[Hint: Check reference [6].]*

b. Compute and plot the radial drift of the particle (average radial displacement in one time step) as a function of the radius ($x^2 + y^2$). Can you find a value of the delay for which the radial drift vanishes? *[Hint: Check reference [6].]*

10.3 Multiple robots with sensory delay

We will now extend our observations for a single robot to multiple robots. In this case, instead of having an externally-applied light field, each robot will emit its own light field, which is then measured by the other robots. Each robot will measure the cumulative light intensity that is created by the other robots, while disregarding its own light. As for the case of a single robot, the multiple robots also adjust their speed based on their measured light intensity with a delay of δ.

We consider n robots with positions (x_i, y_i), $i = 1, \dots, n$. Assuming that each robot emits a light field with a Gaussian profile, the total light intensity reaching each robot can be calculated as follows:

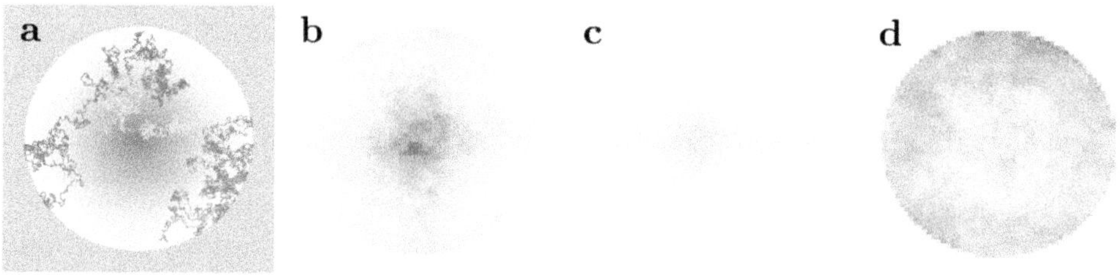

Figure 10.5. Robot in a circular well. (a) An autonomous robot that moves according to equation (10.1) is placed in a Gaussian light field confined within a circular well with solid boundaries. The distribution of the robot's trajectory can be tuned by controlling δ. Sample trajectories are shown with no sensory delay (blue line, $\delta = 0$), positive sensory delay (yellow line, $\delta = 5\tau$), and negative sensory delay (green line, $\delta = 5\tau$). (b) With $\delta = 0$, the robot spends most of its time at intermediate radius values. (c) With a positive sensory delay ($\delta = 5\tau$), the robot spends most of its time in the highest-intensity (i.e., lowest-speed) region. (d) With a negative sensory delay ($\delta = -5\tau$), the robot spends most of its time in the lowest-intensity (highest-speed) region. The parameters used for this simulation are $R = 2$ m, $I_0 = 1$ W, $r_0 = 1$ m, $V_0 = 1$ m s^{-1}, and $V_\infty = 0.1$ m s^{-1}.

$$I_i(t) = \sum_{j \neq i} I_0 \exp\left[-\frac{(x_j - x_i)^2 + (y_j - y_i)^2}{r_0^2}\right], \tag{10.4}$$

where r_0 is the decaying length scale of the signal and I_0 is the intensity of the light emitted by each robot. Since individual robots are attracted to the high (low) intensity region for positive (negative) values of sensory delay, multiple light-emitting robots cluster together for positive delay (as shown in figures 10.6(d)–(f)), while they diverge from each other for negative delay (as shown in figures 10.6(g)–(i)).

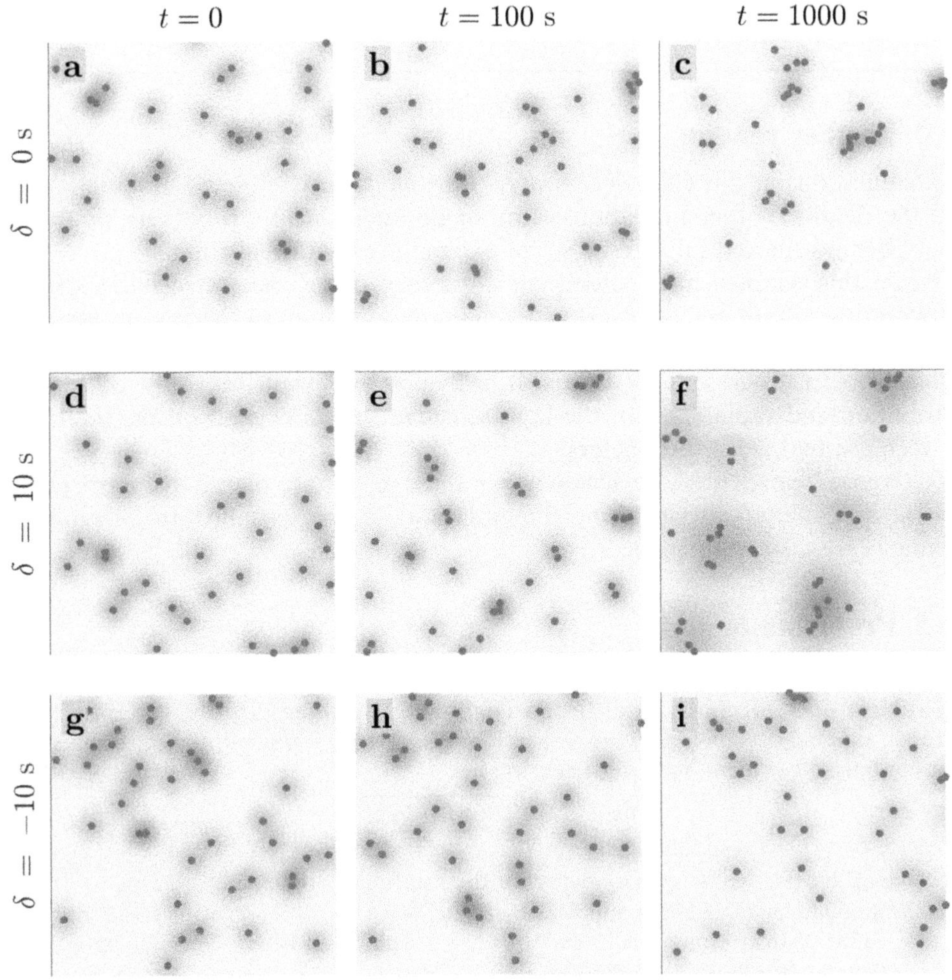

Figure 10.6. Tuning the collective behavior of robots using sensory delay. (a)–(c) Snapshots of 40 autonomous light-sensitive robots moving in an arena with no sensory delay at (a) $t = 0$ s, (b) 100 s, and (c) $t = 1000$ s. The robots tend to stay close to each other. (d)–(f) With a positive sensory delay, $\delta = +10$ s, the robots get closer over time and form clusters. (g)–(i) With a negative sensory delay, $\delta = -10$ s, the robots move away from each other and spread around the arena. The parameters used for this simulation are $R = 2$ m, $I_0 = 5$ W, $r_0 = 1$ m, $V_0 = 1$ m s^{-1}, and $V_\infty = 0.1$ m s^{-1}.

Exercise 10.4. Multiple robots with sensory delay. Simulate $n = 50$ robots in a square box with sides of length $L = 10$ with periodic boundary conditions. Calculate the light field measured by each robot using equation (10.4) and adjust the robot's speed according to equation (10.2). Use $\tau = 2$ s, $r_0 = 1$ m, and $I_0 = 5$ W. Place the robots randomly and start the simulation.

a. Simulate this multiple robot system and observe its behavior. Show that the robots tend to stay close to each other when there is no delay, as shown in figures 10.6(a)–(c).

b. Show that the robots come closer together when there is a positive delay and form clusters, as shown in figures 10.6(d)–(f).

c. Show that the robots move away from each other and explore the arena when there is a negative delay, as shown in figures 10.6(g)–(i).

10.4 Further reading

The simulations in this chapter are largely inspired by reference [6], where you can find the details of the simulations with single and multiple robots as well as the theory behind these results and a set of experiments with real robots.

While this chapter has focused on a sensory delay that affects the speed of autonomous robots, reference [7] generalizes these results to the case in which the robot's rotational diffusion coefficient also depends on the measured light intensity and is subject to a sensory delay. It provides an extensive study of how the distribution and radial drift of the light-sensitive robots depend on the interaction between the two delay parameters.

Reference [8] provides a review in which these results are placed in a more general context and related to stochastic integrals and to the so-called Itô-Stratonovich dilemma (see also chapter 7).

10.5 Problems

Problem 10.1. Rotational diffusion coefficient with sensory delay. Consider a robot that adjusts not only its speed but also its rotational diffusion coefficient according to the measured light intensity, i.e.,

$$\frac{d\phi(t)}{dt} = \sqrt{\frac{2R(I(t - \delta_R))}{\tau}}\, w_\phi,$$

where δ_R represents the sensory delay for the rotational diffusion coefficient.

a. Simulate a single agent and observe its behavior for different values of δ and δ_R. Study its distribution and radial drift in a circular well with a Gaussian light field (similar to that used in exercise 10.3.)

b. Simulate multiple robots and study the system with different values of δ and δ_R.

c. How do these results depend on the ratio between δ and δ_R?

Problem 10.2. Alternative speed functions. Modify the dependence of the robot's speed on the measured light intensity using the following formula:

$$V(I) = V_0 - (V_0 - V_{\mathrm{inf}})e-(I - I_c)2,$$

where I_c is the light intensity at which the robot's speed reaches its minimum. Perform single and multiple robot simulations and study the system with different values of I_c. Can you control the characteristic size of the robot clusters in this way?

Problem 10.3. Stable clusters. Simulate the formation of clusters by introducing a delay into a group of robots, as in exercise 10.4. Can you stabilize the cluster size by periodically modulating the intensity of the light emitted by the robots?

Problem 10.4. Search and rescue in a complex environment. Consider some targets dispersed in a complex environment (i.e., an environment where several obstacles are present). Find a strategy for employing a group of autonomous robots to explore this environment, collect the targets, and finally gather at the initial position. *[Hint: For example, you can build on the fact that a negative sensory delay disperses the robots and a positive one regathers them.]*

Problem 10.5. Escaping from a labyrinth. Consider a labyrinth and a group of light-emitting robots responding to each other's signals with a delay. Study how they can optimize their chances of escaping from the labyrinth by playing with the delay.

10.6 Challenges

Challenge 10.1. Delayed Vicsek model. Introduce a delay into the Vicsek model (see chapter 8) and determine how its behavior changes. *[Hint: Check reference [9].]*

Challenge 10.2. Delayed Lotka–Volterra model. Simulate an agent-based version of the Lotka–Volterra prey–predator model and introduce delays into the interactions between prey and predators. Study how the success of prey and predators changes as a function of these delays. *[Hint: See chapter 14 for a discussion of the Lotka–Volterra model.]*

Challenge 10.6.3. Chemotactic signals. Study the case in which the signaling between robots occurs through chemotactic molecules diffusing from each robot instead of light (i.e., the case in which the signal itself propagates with a certain propagation speed and decays over time). Connect your observations for this situation to the concept of *quorum sensing*.

References

[1] Davids A 2002 Urban search and rescue robots: from tragedy to technology *IEEE Intell. Syst.* **17** 81–3

[2] Shapiro J A 1998 Thinking about bacterial populations as multicellular organisms *Annu. Rev. Microbiol.* **52** 81–104

[3] Berg H C 2008 *E. coli in Motion* (Springer: Berlin)

[4] Yu W and Egbelu P J 2008 Scheduling of inbound and outbound trucks in cross docking systems with temporary storage *Eur. J. Oper. Res.* **184** 377–96

[5] Moussaïd M *et al* 2009 Experimental study of the behavioural mechanisms underlying self-organization in human crowds *Proc. Royal Soc.* B **276** 2755–62

[6] Mijalkov M, McDaniel A, Wehr J and Volpe G 2016 Engineering sensorial delay to control phototaxis and emergent collective behaviors *Phys. Rev.* X **6** 011008

[7] Leyman M, Ogemark F, Wehr J and Volpe G 2018 Tuning phototactic robots with sensorial delays *Phys. Rev.* E **98** 052606

[8] Volpe G and Wehr J 2016 Effective drifts in dynamical systems with multiplicative noise: a review of recent progress *Rep. Prog. Phys.* **79** 053901

[9] Piwowarczyk R, Selin M, Ihle T and Volpe G 2019 Influence of sensorial delay on clustering and swarming *Phys. Rev.* E **100** 012607

IOP Publishing

Simulation of Complex Systems

Aykut Argun, Agnese Callegari and Giovanni Volpe

Chapter 11

Disease spreading

Infectious diseases that spread in human populations represent a growing global challenge, as we have experienced with the recent coronavirus disease (COVID-19) pandemic (figure 11.1). An individual affected by an infectious disease can transmit it to several other individuals, who can, in turn, become infected and transmit the illness to many others, in an almost exponential progression. When the disease is severe and hospitalization or intensive care is needed, an uncontrolled spread can severely overwhelm any healthcare system. The logistic burden entails the practical impossibility of dispensing the required care to each affected individual, leading to

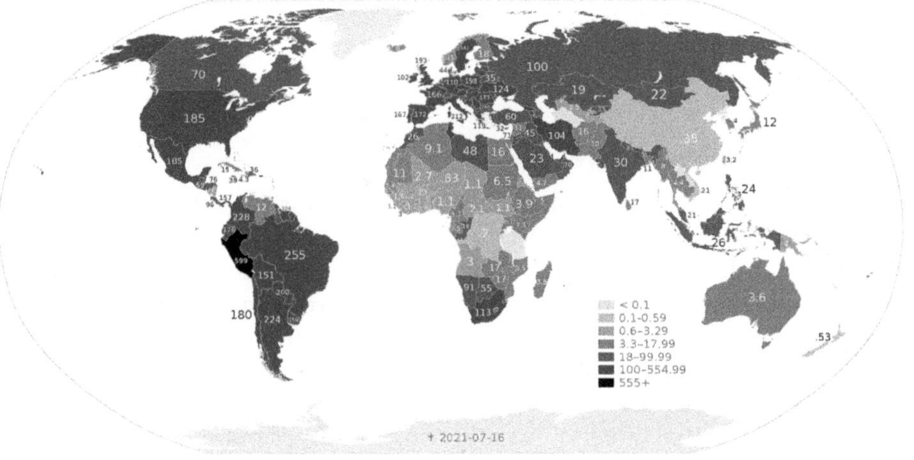

Figure 11.1. World map of the total deaths per 100,000 inhabitants due to the coronavirus disease pandemic. The COVID-19 pandemic rapidly became one of the biggest global problems during 2020 and 2021. By July 2021, the virus had already spread to hundreds of millions of people and killed several millions in all countries. The total deaths per capita (updated to July 16, 2021) are shown in the figure. *Source: Wikipedia: Dan Polansky (CC BY-SA 4.0).* https://commons.wikimedia.org/wiki/File:COVID-19_Outbreak_World_Map_Total_Deaths_per_Capita.svg

an increase in the number of deaths. This poses several ethical questions and puts considerable stress on healthcare providers.

To choose the most effective policy for the mitigation and containment of outbreaks, it is very important to understand the mechanisms responsible for the spread of disease. Infectious diseases often spread in a society through individual contact. This can take place through direct contact, via a sneeze or cough, or by exchange of bodily fluids. The dynamics of an infectious disease depends on mobility, infection probability, and recovery speed. The susceptible–infected–recovered model (*SIR model*), introduced by W. O. Kermack and A. G. McKendrick in 1927 [1], which built on previous work by Sir Roland Ross and Hilda Hudson [2–4]) studied the disease transmission mathematically, providing a series of ordinary differential equations (ODEs) that capture its essential properties. Many studies have built on this model, shining light on various aspects of disease transmission. Agent-based models are a natural next step from the simple ODE approach. They have become an important tool in understanding disease dynamics and developing countermeasures.

In this chapter, we describe a simple agent-based SIR model that has been used in epidemiology to model the spread of various infectious diseases [5]. We start with the basic principles of the SIR model in order to numerically simulate the transmission of diseases in a finite environment using agents. We then introduce further parameters that allow us to develop more realistic versions of the SIR model, such as incubation time, mortality, and hospitalization. Finally, we analyze how lockdowns can contain the transmission of a disease.

Example codes: Example Python scripts related to this chapter can be found at: https://github.com/softmatterlab/SOCS/tree/main/Chapter_11_Disease_Spreading. Readers are welcome to participate in the discussions related to this chapter on: https://github.com/softmatterlab/SOCS/discussions/20.

11.1 The agent-based SIR model

We first consider the three-compartment model known as SIR, in which each individual is either Susceptible (S) to the disease, Infected (I), or has Recovered (R) and is immune. Infected individuals infect the susceptible they meet at a rate of β and recover at a rate of γ. In the original model [1], the problem is solved using an ODE approach while considering that the infection is a continuous process. In such a version of the model, only the ratio $k = \beta/\gamma$ matters for the behavior of the model. With an agent-based model, it is also possible to examine what happens when we take into account spatial effects. The following is a description of the series of steps required to implement the agent-based SIR model:

1. On a finite arena (e.g., a square grid lattice), at $t = 0$, initialize N agents at random locations and make a small fraction of the agents infected.
2. Model the movements of the agents as random walks on the arena: at every time step, each agent either remains still with some probability 1∸, or moves

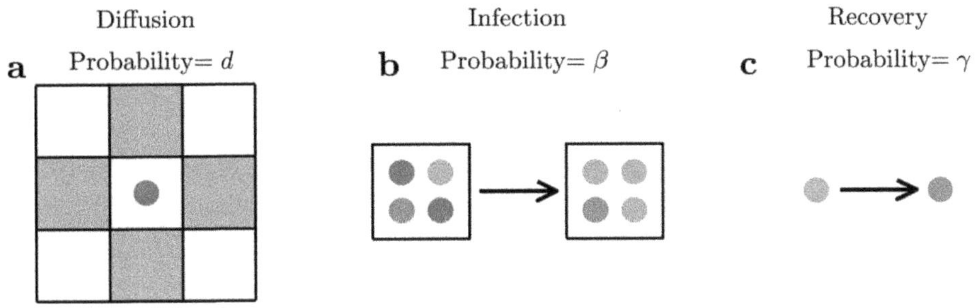

Figure 11.2. Agent-based SIR model. At each time step: (a) every agent (blue dot) moves to a random neighboring site (its von Neumann neighborhood, orange boxes) with a probability of d. (b) Every infected agent (orange dot) has probability β of infecting all the susceptible agents (blue dots) that are in the same site. The recovered agents (green dot) remain unaffected by the infection, as they are immune. **c.** Every infected agent (orange dot) recovers (green dot) with a probability of γ.

to a random neighboring tile (e.g., their von Neumann neighborhood, the orange boxes in figure 11.2(a)) with a probability of d, where d denotes the diffusion rate.

3. For every infected agent, check if there are any susceptible agents in the same lattice site. At every time step, each infected agent should have a probability β of infecting all susceptible individuals at its current site, as shown in figure 11.2(b).
4. At every time step, each infected agent has a probability of recovery of γ, as shown in figure 11.2(c).
5. The disease dynamics evolves until the number of infected agents reaches zero. This condition can be used as a stopping criterion for the simulation.

A screenshot of an agent-based SIR simulation is shown in figure 11.3(a). Every simulation starts with zero recovered agents and ends with zero infected agents, as shown in figure 11.3(b). In the first exercise, we will simulate this simple agent-based SIR model.

Exercise 11.1. Simulation of the SIR model. Implement the basic model as explained above and visualize it. Start with just a few agents and make sure they perform the random walk, infection, and recovery correctly. Then, test a small number of agents to check that the disease dynamics seems reasonable. Choose parameters $\gamma = 0.01$ and $d = 0.8$. Use β as the varying parameter.

a. Scale the simulation up (e.g., 1000 agents on a 100×100 lattice, as shown in figure 11.3(a)). Use an initial infection rate of 1%.

b. Show that the model has two regimes, i.e., that there are parameters for which the disease spreads to a large proportion of the population and values for which it does not. *[Hint: For fixed values of γ and d, find the critical β value that makes the disease spread to a majority of the population.]*

c. Plot the numbers of susceptible (S), infected (I), and recovered (R) agents as functions of time (as shown in figure 11.3(b)).

d. Repeat the simulations for various values of β and γ. Comment on your results.

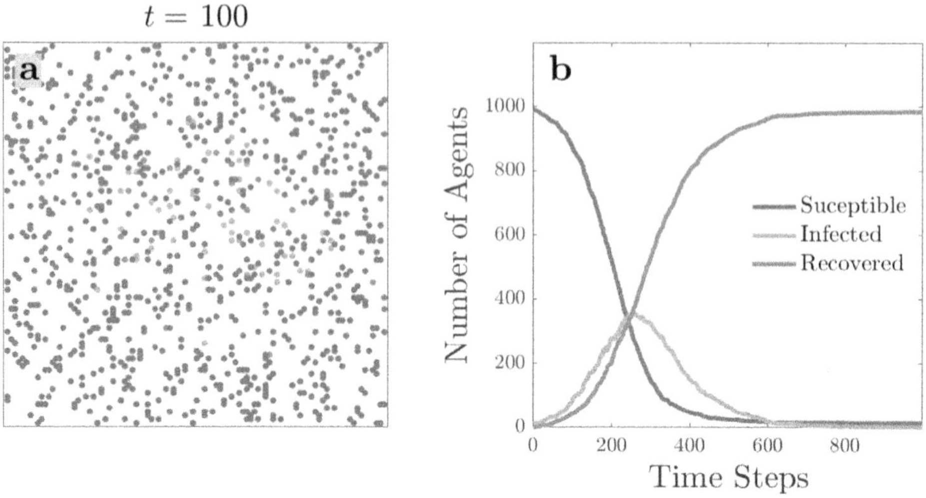

Figure 11.3. Modeling disease transmission on a square lattice. (a) Screenshot of SIR agents on a lattice after 100 time steps. The blue dots represent susceptible agents, orange dots represent infected agents, and green dots represent recovered agents. (b) Numbers of susceptible (blue), infected (orange), and recovered (green) individuals as functions of the time step in the agent-based SIR model. The parameters used for this simulation are $d = 0.8$, $\beta = 0.6$, and $\gamma = 0.01$. At $t = 0$, 1% of the total number of agents (10 out of 1000) are initialized as infected.

11.2 Disease transmission as a function of the infection rate

In the SIR model, the reach of a pandemic can be measured by the final number of recovered agents, which is denoted by R_∞. In fact, at the end of a simulation, the population is divided into agents that have never been infected and agents that have been infected and recovered. Naturally, R_∞ depends strongly on the infection rate β. This dependence is demonstrated in figure 11.4 for $\gamma = 0.01$ (blue dots) and $\gamma = 0.02$ (green dots). Interestingly, in the original ODE model, the value of R_∞ only depends on the ratio β/γ. However, when we take into account of the movement of each individual within the agent-based model, this is no longer true.

Exercise 11.2. Dependence of the final number of recovered agents on the infection rate. Simulate the agent-based SIR model with $d = 0.8$ and $N = 1000$. Use a 1% initial infection rate. For all the questions in this exercise, repeat the simulation several times to obtain the average R_∞.

a. Set the recovery rate $\gamma = 0.01$ and plot R_∞ as a function of β, as shown in figure 11.4.

b. Set $\gamma = 0.02$ and compare the results to those for part (a). Convert the figure into R_∞ vs β/γ for both γ values, and show that the results do not only depend on β/γ but also on the value of the infection rate β.

c. Plot the phase diagram of R_∞ as a function of β and β/γ. Show that the output depends not only on β/γ but also on β itself, as shown in figure 11.5.

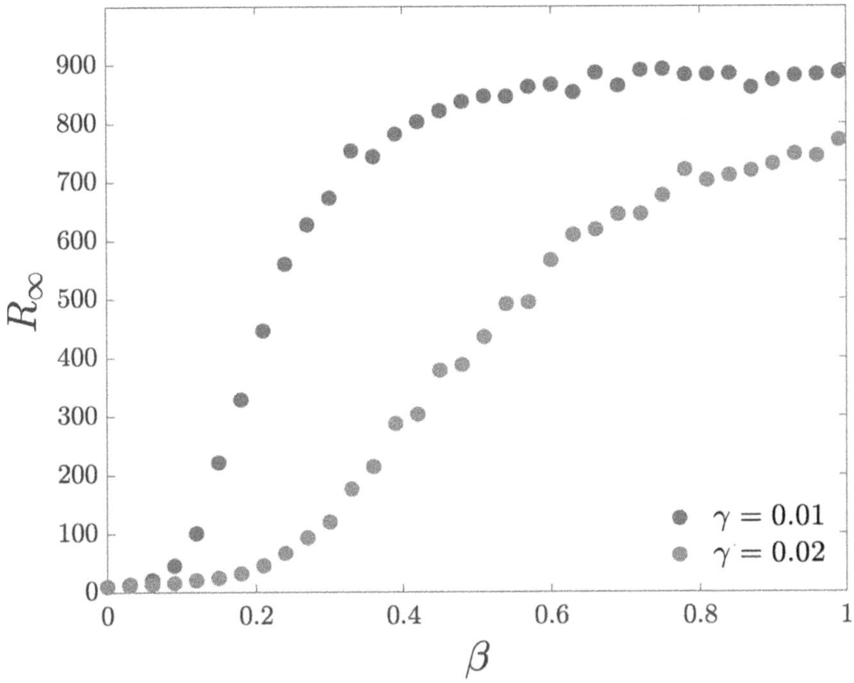

Figure 11.4. Final number of recovered agents as a function of the infection rate. The average number of recovered agents at the end of the simulation (R_∞) as a function of the infection probability β for $\gamma = 0.01$ (blue dots) and for $\gamma = 0.02$ (green dots). The results for each dot are averaged over 100 repetitions of the simulation ($d = 0.8$, $N = 1000$, and a 1% initial infection rate).

11.3 Extended SIR models

Although it is a very powerful minimalist model, the SIR model does not fully describe the underlying behavior of every possible pandemic. In fact, there are many assumptions underlying the SIR model, which might not reflect real-world conditions. A list of the main oversimplifications that the agent-based SIR model makes is as follows:

- It assumes that the infection is immediate and that people can infect others as soon as they become infected. However, in real life, there is a delay between the time at which an individual is exposed to the virus and the time at which they develop the infection themselves and become capable of infecting others (in fact, these latter two times might not even coincide).
- It assumes that all cases are asymptomatic, which means that the mobility of infected agents does not change at all. This is not true, as, in real life, sick individuals tend to move less.
- It assumes that immunity is permanent. However, immunity often reduces over time after recovery, and reinfection is often possible.

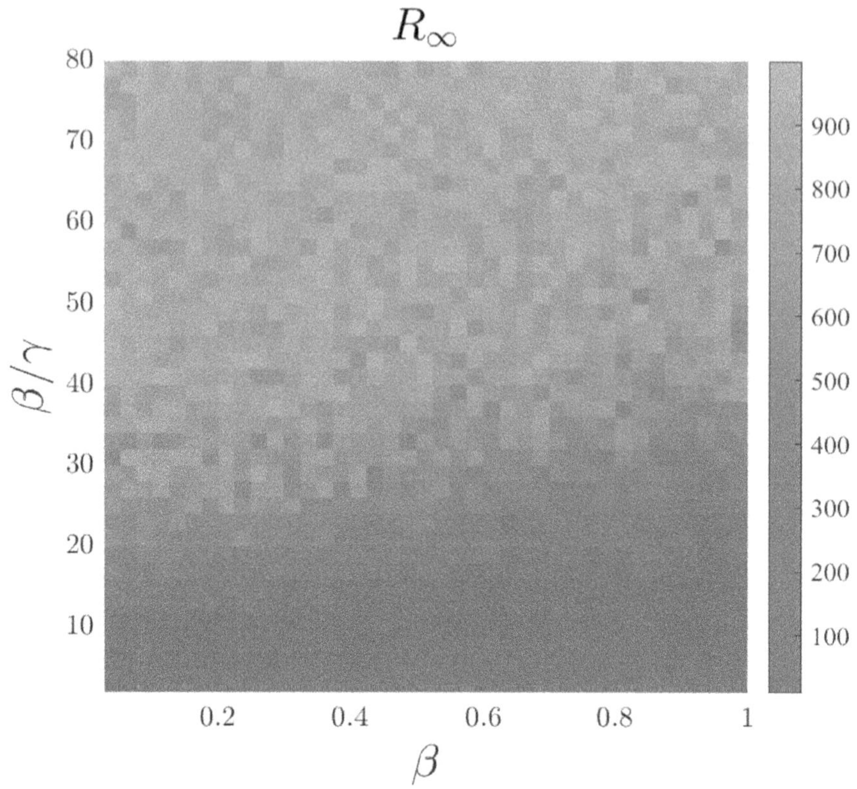

Figure 11.5. Final number of recovered agents as a function of β and β/γ. The average number of recovered agents at the end of the simulation (R_∞) as a function of the infection probability β and β/γ. The results are averaged over three repetitions of the simulation ($d = 0.8$, $N = 1000$, and a 1% initial infection rate).

- It assumes that the disease does not cause hospitalization or death. This may not always be the case. In fact, many global pandemics, including the COVID-19 pandemic, have caused millions of deaths.

Depending on the purpose of the simulation model, these assumptions can be relaxed by implementing extended versions of the SIR model. Although these extended SIR models increase the computational complexity, they capture a wider range of outputs for the transmission of a disease. Therefore, to make our SIR model more realistic, we are now going to make it more complex and relax some of the assumptions mentioned above.

Exercise 11.3. SIR model with mortality. Simulate the agent-based SIR model, adding a probability μ that infected agents die at each time step. Denote the number of dead agents at the end of the simulation by D_∞. At each time step, simulate infection, recovery, and death. Start with an initial infected agent rate of 1%.

a. Show that the number of final casualties does not monotonically depend on μ. Show that for both very low and very high values of μ, the final number of casualties is minimal. Demonstrate the range of μ for which the pandemic becomes most severe.

b. Analyze and plot the dependence of D_∞ on μ for different values of β and γ. Play with the parameters and identify the most dangerous value of μ for different values of β and γ.

Exercise 11.4. SIR model with temporary immunity. Simulate the agent-based SIR model, adding a probability α that recovered agents become susceptible again.

a. Simulate this model and plot the number of susceptible and infected agents as a function of time. What happens at long times?

b. Play with the parameters α, β and γ. For which parameters does the disease die out? And for which ones does it become endemic? Does this depend on the number and distribution of the initial infected agents?

11.4 Lockdown strategies

If a disease with a high mortality rate gets out of control and spreads to the general public, it can disrupt society. For example, the Antonine plague destroyed the social fabric of the Roman Empire during the 2nd century AD, eventually causing 30% of its population to lose their lives. Another example is the bubonic plague, which killed more than 30% of the European population in the fourteenth century. Because of the ever-increasing population density and global travel, major pandemics in modern ages, such as the Spanish flu and the COVID-19 pandemic, have spread worldwide.

An ideal strategy for quashing a pandemic is to identify and isolate all infected individuals. However, this is often logistically impossible due to the difficulties of making a timely diagnosis. Therefore, lockdowns have often been imposed as a practical strategy in order to minimize the overall damage caused by these pandemics. However, lockdowns themselves also entail serious social, financial, and psychological damage. Therefore, it is very important to identify the optimum strategies for imposing effective partial lockdowns during a pandemic.

In the next exercise, we will implement a lockdown strategy that will limit the spread of the disease. To realistically model this behavior, we will only be allowed to hold the lockdown in place for a limited amount of time.

Exercise 11.5. Simulation of the simple SIR model with a lockdown. Simulate the simple SIR model with the parameters $\gamma = 0.01$, $\beta = 0.5$, and $d = 1$, starting with 1000 agents and an initial 1% of infected agents. Assume that we are only allowed to operate a lockdown of 200 steps.

a. Start the lockdown at different time steps and observe the resulting R_∞. Plot the number of susceptible, infected, and recovered individuals as a function of time for each case and comment on your results. Decrease the diffusion coefficient by one order of magnitude to $d = 0.1$ while the lockdown is in place.

b. Calculate the optimum lockdown time and identify the best strategy, i.e., the one that results in the minimum value of R_∞.

11.5 Further reading

Reference [5] is an excellent source with which to study the SIR model based on ODE; it provides both the theoretical relations and their application to historical data. Reference [6] is another article that presents three different ODE-based epidemiological models. Reference [7] is a great book that explains the theory of the standard epidemiological models in detail and presents case studies that analyze data from real outbreaks.

Reference [8] is an excellent study, which shows that targeted lockdown strategies reduce the spread of disease. The authors show that the optimal strategy can be found by targeting age and risk groups differently.

Reference [9] uses techniques from optimization theory and machine learning to conduct optimizations of alternative disease policies.

Reference [10] demonstrates a machine-learning approach to the analysis of initial confirmed cases to provide strategies for targeted testing and lockdown. These machine-learning-based strategies are shown to contain an outbreak more effectively than the standard approaches.

11.6 Problems

Problem 11.1. Dependence on the diffusion coefficient. Implement the simple SIR simulation for 1000 agents, as done in exercise 11.1.

- **a.** Calculate and plot R_∞ as a function of d.
- **b.** Repeat the same simulations with a Moore neighborhood and compare these results to those for the von Neumann neighborhood.

Problem 11.2. Phase diagram. Implement the SIR simulation for 1000 agents, as done in exercise 11.1. Calculate R_∞ for a wide range of values for both β and γ. Calculate the dependence of R_∞ on β and β/γ for different lattice sizes. Show that the dependence of R_∞ on β (figure 11.5) is stronger for larger lattice sizes.

Problem 11.3. SIR model with incubation time. Simulate the SIR model with infected agents that get sick after a certain incubation time δ (i.e., the delay between the time at which an agent becomes infected and the time when it can infect others). Analyze the spread of the pandemic as a function of this new parameter δ.

Problem 11.4. Maximum number of infected agents. Another important parameter for the SIR model is the maximum number of infected agents. This is because the healthcare system may be overwhelmed by an excessive number of infected people at the same time, so that people may not be able to receive appropriate medical care.

- **a.** Simulate the agent-based SIR model for 1000 agents. Let I_{max} be the maximum number of infected agents during a simulation. Calculate and plot I_{max} as a function of β. Repeat and average the simulations to obtain a smooth plot.
- **b.** Plot the dependence of I_{max} on β and β/γ, as in figure 11.5. Discuss your results.

Problem 11.5. SIR with hospitalization. Assume that in the SIR model, some of the agents that become infected are either hospitalized or stay at home more often than the healthy agents. In the simulation, this can be effectively represented by a coefficient $s < 1$ that modulates the diffusion probability of the infected agents, so that their diffusion probability is reduced to sd. Simulate the SIR model with different values of s and discuss the results.

11.7 Challenges

Challenge 11.1. Agent-based SIR with commuter agents. Modify the SIR simulation so that each agent has home and work locations; each agent goes to work in the morning and comes back home in the evening. Make the path stochastic, such that each agent follows a slightly different path each time. How does this change the behavior of the model? You can then make your model even more complex by introducing other locations (e.g., schools, hospitals, parks), groups of agents (e.g., workers, students, retirees), and different days (e.g., work days and holidays). Furthermore, the parameters β and γ can also be modulated as a function of the location. Finally, you can also introduce a topology of the environment that maps that of a real city. With these modifications, the SIR model becomes an extremely powerful predictive tool with which to forecast the spread of a disease in a real environment.

Challenge 11.2. Use of neural networks to develop targeted testing and lockdown strategies. Write a program using neural networks to develop strategies for containing a pandemic by targeted testing and lockdown. Use simulated data to train your network. *[Hint: Check reference [10].]*

References

[1] Kermack W O, McKendrick A G and Walker G T 1927 A contribution to the mathematical theory of epidemics *Proc. R. Soc. Lond.* A **115** 700–21

[2] Ross R and Hudson H P 1916 An application of the theory of probabilities to the study of *a priori* pathometry. Part I *Proc. R. Soc. Lond.* A **92** 204–30

[3] Ross R and Hudson H P 1917 An application of the theory of probabilities to the study of *a priori* pathometry. Part II *Proc. R. Soc. Lond.* A **93** 212–25

[4] Ross R and Hudson H P 1917 An application of the theory of probabilities to the study of *a priori* pathometry. Part III *Proc. R. Soc. Lond.* A **93** 225–40

[5] Weiss H 2013 The SIR model and the foundations of public health *MATerials MATemàtics* **2013** 3

[6] Hethcote H W 1989 Three basic epidemiological models In *Appl. Math. Ecol.* (Berlin: Springer) pp 119–44

[7] Allen L J S, Brauer F, Van den Driessche P and Wu J 2008 *Mathematical Epidemiology* (Berlin: Springer)

[8] Acemoglu D, Chernozhukov V, Werning I and Whinston M D 2020 *Optimal targeted lockdowns in a multi-group SIR model 27102* National Bureau of Economic Research Working Paper Series https://nber.org/papers/w27102

[9] Navascués M, Budroni C and Guryanova Y 2021 Disease control as an optimization problem *Plos One* **16** 1–32

[10] Natali L, Helgadottir S, Maragò O M and Volpe G 2021 Improving epidemic testing and containment strategies using machine learning *Mach. Learn.: Sci. Technol.* **2** 035007

IOP Publishing

Simulation of Complex Systems

Aykut Argun, Agnese Callegari and Giovanni Volpe

Chapter 12

Network models

Networks (or *graphs*) can describe a wide range of systems in countless disciplines, including physics, biology, finance, ecology, sociology, and the neurosciences. A historical example of a puzzle that was solved using graph theory is the *seven bridges of Königsberg*. Königsberg is a small central European town with four landmasses connected by seven

Figure 12.1. The Seven Bridges of Königsberg. A map of the central European city of Königsberg in the 18th century with its seven bridges. Is it possible to find a path that passes through each bridge once and only once, ending at the initial position? This puzzle was solved in 1736 by the great mathematician Leonhard Euler, who used graph theory to prove that such a path does not exist (can you prove it?). *Source: Wikipedia: Encik Tekateki (Public Domain).* https://commons.wikimedia.org/wiki/File:Konigsberg_Bridge.png

bridges, as shown in figure 12.1. In the 18th century, people were puzzled by the question of whether it was possible to find a path that passed through each bridge of the town once and only once, ending at the initial position. This question was then answered in 1736 by the great mathematician Leonhard Euler, who managed to prove that such a path cannot exist by modeling this problem using a graph. In fact, by solving this problem, Euler created the new mathematical fields of *graph theory* and *topology*.

Network models have been intensely studied to describe the properties of graphs. One of the most common mathematical ways to understand a network is to represent it using an *adjacency matrix*, which is a matrix in which each entry describes the connection between two nodes [1]. Depending on the kinds of connection between nodes, adjacency matrices can be *binary* (a connection either exists or not) or *weighted* (a connection has a numerical value). Furthermore, they can be *symmetric* (for *directed graphs*) or *asymmetric* (for *undirected graphs*). There are many ways to initialize and grow networks to model systems with different properties.

In this chapter, we start by defining the adjacency matrix. We then implement different network initialization methods, such as the *Erdős–Rényi random graph*, the *Watts–Strogatz small-world model*, and the *Albert–Barabási preferential-growth model*. We also learn how to analyze networks using graph metrics such as the *average path length*, the *diameter*, and the *clustering coefficient*.

Example codes: Example Python scripts related to this chapter can be found at: https://github.com/softmatterlab/SOCS/tree/main/Chapter_12_Network_Models. Readers are welcome to participate in the discussions related to this chapter at: https://github.com/softmatterlab/SOCS/discussions/21.

12.1 The adjacency matrix

A graph is made of a set of *nodes* connected by *edges*. Depending on the specific graph, the nodes and edges can represent, e.g., cities and the physical distances between them, brain regions and the number of white matter fibers connecting them, or servers and the internet bandwidth between them. If the network is *binary*, the edge between any two nodes is either present (1) or absent (0)—such as in a graph in which the nodes are airports and the edges represent the existence of flights between them. If the network is *weighted*, each edge acquires a real numerical value—such as in a graph in which the nodes are cities and the edges represent distance between them. Furthermore, if the network is *directed*, the connections have directionality (e.g., node j can be connected to node k, but not vice versa)—such as those of a food web, where the nodes represent species and the edges indicates who eats whom. If the network is *undirected*, its connections do not have preferential directionality (i.e., if node j is connected to node k, then node k is automatically connected to node j)—for example, in the graph of the distance between cities.

A graph can be represented by its *adjacency matrix*, which we are going to denote by A. If a graph has n nodes, this means that A is going to be an $n \times n$ matrix, where

A_{ij} represents the connection between the nodes i and j. If the network is symmetric, then $A_{ij} = A_{ji}$. If the graph is binary, then each element in A is either one or zero. For a binary graph, the adjacency matrix has the following properties:

- The sum of the adjacency matrix along one dimension, $\sum_i A_{ij}$, gives the *degree* of a node (i.e., its number of connections).
- The t-power of the adjacency matrix, A^t, gives the number of t-step paths from node i to node j (can you prove this?).

In the following exercises, we are going to generate some graphs using various popular network initialization and growth models.

Exercise 12.1. The Erdős–Rényi random graph. This model is symmetric, binary, and has two parameters, n and p. It consists of n nodes, and each of the $n(n-1)/2$ possible edges is present with a probability p. In order to generate the adjacency matrix of such an Erdős–Rényi random graph:

1. Generate an empty adjacency matrix with zeros which has a size of $n \times n$.
2. In a double for-loop, change A_{ij} and A_{ji} (for $j > i$) to one with probability of p.

a. Generate a Erdős–Rényi random graph and plot it, as shown in figure 12.2(a). Visualize a few examples for different values of n and p.

b. Show that the probability that a generic node has degree k is

$$P(k) = \binom{n-1}{k} p^k \left(1 - p\right)^{n-1-k}, \tag{12.1}$$

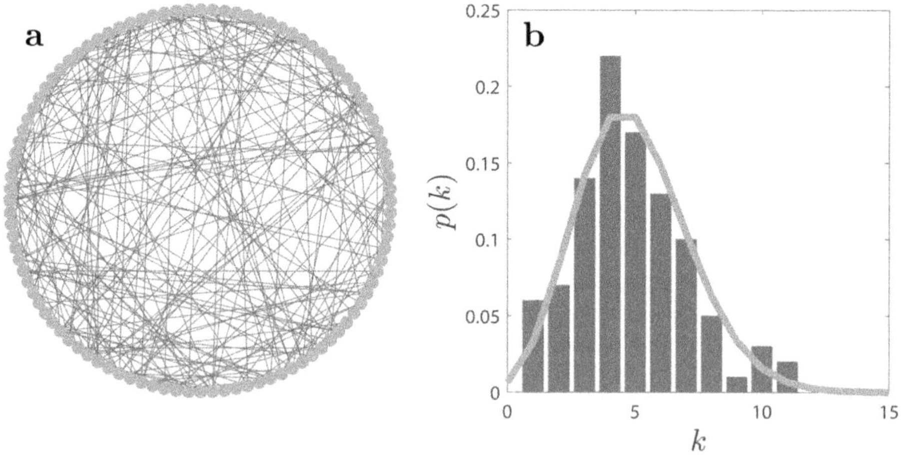

Figure 12.2. The Erdős–Rényi random graph. (a) An example of an Erdős–Rényi random graph with $n = 100$ nodes (orange dots) connected with a probability $p = 0.05$ (blue lines represent the edges). (b) The resulting degree distribution (blue bars) is shown and compared to the theoretical distribution (orange line, see exercise 12.1).

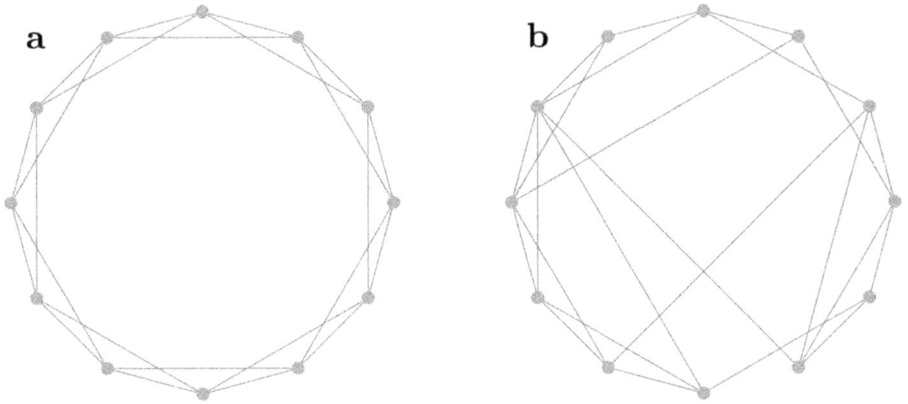

Figure 12.3. The Watts–Strogatz small-world graph. (a) An example of a Watts–Strogatz graph ($c = 4$) with no rewiring ($p = 0$). (b) The same graph, but with a probability of rewiring of $p = 0.2$.

as shown in figure 12.2(b). Repeat this for different values of n and k to further verify the formula.

c. Show that the degree distribution converges to a Gaussian for large n and small p.

Exercise 12.2. The Watts–Strogatz small-world model. This is also a symmetric and binary model. This model has three parameters, n, c, and p; c should be even. It consists of n nodes placed in a circle, each connected to its c nearest neighbors (so that $c = 2$ gives an ordinary circle) with additional random connections. To generate this network, we follow these steps:

1. Generate an empty adjacency matrix with zeros which has a size of $n \times n$.
2. For every i and j, make $A_{ij} = A_{ji} = 1$ if $|i - j|$ ranges from 1 to $c/2$ (the norm should be calculated with periodic boundary conditions).
3. In a double for-loop, randomly rewire each connection with a probability of p.

a. With $n = 20$, $p = 0$, and $c = 2$, generate a Watts–Strogatz small-world graph and plot it, as shown in figure 12.3(a). Repeat for $c = 4$ and $c = 8$.

b. Repeat for $p \neq 0$ and visualize the difference, as in figure 12.3(b).

Exercise 12.3. The Albert–Barabási preferential-growth model. This is a model of network formation that has one important parameter, m. It starts with a configuration of $n_0 \geqslant m$ connected nodes. Then, at each step, for some given number of steps, we add a new node with m new connections. These connections are made with nodes chosen proportionally to their degree, so that the probability of choosing node j is $\Pi(k_j) \propto k_j$. To generate a graph using this model, we follow these steps:

1. Generate an initial fully connected adjacency matrix (filled with ones, except for the diagonal, which should contain zeros) which has a size of $n_0 \times n_0$; choose a small number for n_0 ($n_0 \geqslant m$).
2. In a for-loop with $(n - n_0)$ iterations, add a new node and connect it to m extant nodes chosen randomly using a probability that is proportional to their current degree. Therefore, nodes with higher degrees will be more likely to be connected to new edges.

a. Generate the Albert–Barabási preferential-growth model using various parameters and visualize the output graphs, as shown in figure 12.4(a).

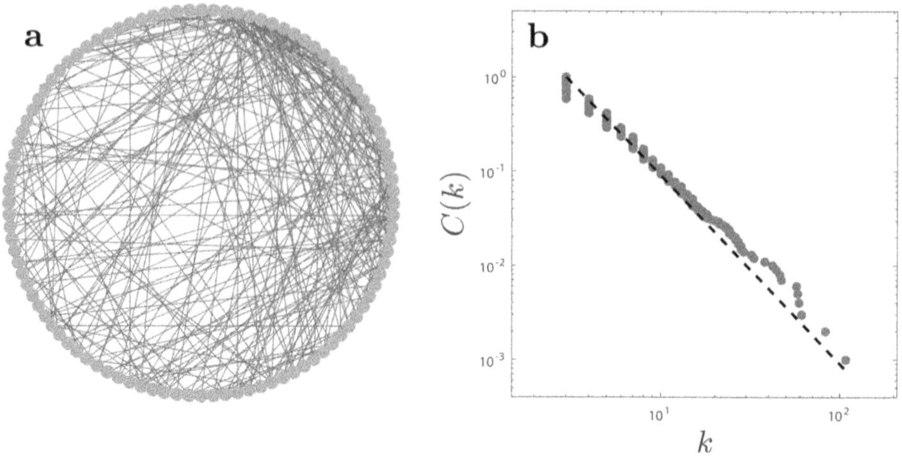

Figure 12.4. The Albert–Barabási preferential-growth graph. (a) An example of a preferential-growth graph with $n = 100$, $m = 3$, and $n_0 = 5$. (b) Inverse cumulative degree distribution for a preferential-growth graph with parameters of $n = 1000$, $m = 3$, and $n_0 = 5$. The dashed line represents the theoretical prediction (see exercise 12.3).

b. Show that the degree distribution follows a power law. Verify this by doing the following: let D_s be the array of sorted degrees in the graph (in descending order) and u be a linear array $[1/n, 2/n, \ldots n/n]$. Plot D_s vs u with both axes in a log scale. This shows the probability that a node has a higher degree than k as a function of k. Compare your results to the theoretical prediction

$$C(k) = m^2 k^{-2}. \tag{12.2}$$

Refer to figure 12.4(b).

12.2 Path length, diameter, and clustering coefficient

The adjacency matrix of a binary graph can be used to calculate various graph metrics, such as the path length, the diameter, and the clustering coefficient. The path length can be used to understand the distance between nodes. The diameter is the longest path length in a graph, and is useful to quantify the linear size of a network. The clustering coefficient is a powerful analysis method used to quantify how the nodes of a graph are clustered together.

Path length and diameter. It is possible to calculate the minimum path length between every two nodes using the adjacency matrix. To do this, we follow these steps:

1. Create a path length matrix L ($n \times n$). Initially, no path length has been found, so we can initialize the elements of L to -1.

2. For increasing values of $t = 1, 2, \ldots$, compute A^t (which denotes the number of t-step paths between nodes, so that, if $(A^t)_{ij} \neq 0$, there is a path from node i to node j) and in a double for-loop over the nodes, if $(A^t)_{ij} \neq 0$ and $L_{ij} = L_{ji} = -1$, set $L_{ij} = L_{ji} = t$. (For computational efficiency, you can keep the power of the adjacency matrix A^t for every t and only multiply by A to increase the power by one.)
3. Continue the while loop until all the path lengths between every pair of nodes have been identified (i.e., there are no off-diagonal elements of L that are still -1). (Consider what the theoretical maximum number of iterations will be.)

Once we have calculated the path length L, we can immediately compute the average path length of a graph, which is the average of the off-diagonal elements of L. The largest value of the path length is the diameter of a graph.

The clustering coefficient. The clustering coefficient of a graph quantifies how the nodes are clustered together. This can represent, for example, the clustering of people in a social network, the connectivity of neurons in the nervous system, or the extent of local aggregation in a public transport network [2]. The clustering coefficient can be obtained using the following expression:

$$C = \frac{\text{Number of closed triples}}{\text{Number of all triples}}.$$

The number of closed triples can be calculated by cubing the adjacency matrix: $(A^3)_{ij}$ gives the number of three-step paths from node i to node j. Therefore, the diagonal elements of A^3 give the number of closed triples for each node. The number of all triples can be straightforwardly calculated from the degree of a node. If node i has a degree of k_i, the number of all triples that include the node is therefore $k_i(k_i - 1)$. Therefore, we can express the clustering coefficient as follows:

$$C = \frac{\sum_i (A^3)_{ii}}{\sum_i k_i(k_i - 1)}. \tag{12.3}$$

12.3 Erdős–Rényi random graphs

The average path length of an Erdős–Rényi random graph depends on the value of the connection probability p. For small values of p, the average path length is theoretically derived as follows [3]:

$$l = \frac{\ln(n) - \gamma}{\ln[p(n-1)]} + \frac{1}{2}, \tag{12.4}$$

where $\gamma = 0.577\,22$ is the Euler–Mascheroni constant. For larger values of p, the path lengths tend to one with a probability of p and to two with a probability of $1 - p$; thus, in this limit, the average path length becomes:

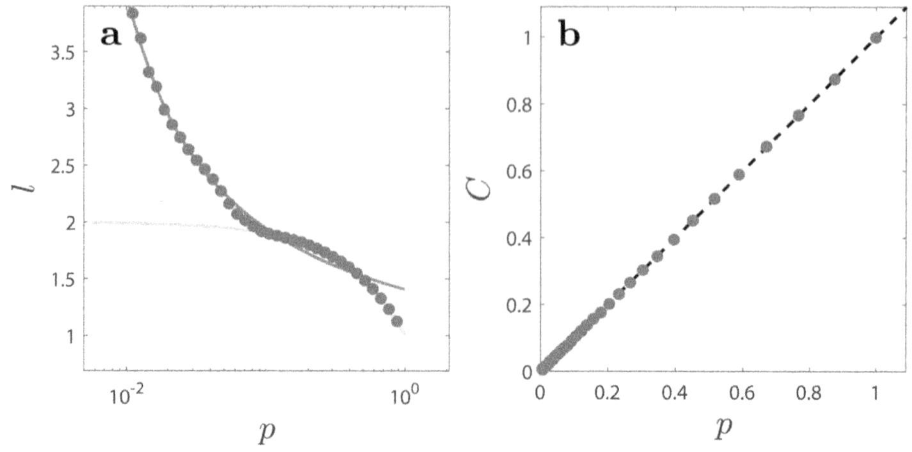

Figure 12.5. Average path length and clustering coefficient of Erdős–Rényi random graphs. (a) The average path length of an Erdős–Rényi random graph (blue dots) as a function of the connection probability p. Theoretical lines are shown for limiting cases when p is small (green line, equation (12.4)) and when p is large (yellow line, equation (12.5)). (b) Clustering coefficient (equation (12.3)) of an Erdős–Rényi random graph (blue dots) as a function of the connection probability p. A graph of size $n = 500$ is used.

$$l = 2 - p. \tag{12.5}$$

This double behavior is shown in figure 12.5(a).

Much more simply, the clustering coefficient of an Erdős–Rényi random graph is equal to p (why?), as shown in figure 12.5(b).

Exercise 12.4. Average path length and clustering coefficient of Erdős–Rényi random graphs. Create large Erdős–Rényi random graphs ($n = 500$) for different p.

a. Calculate the average path length of the graph for each p and plot your results. Verify the theoretical approximations for small and large p, as shown in figure 12.5(a).

b. Calculate the clustering coefficient of the graph as a function of p and show that it is equal to p, as shown in figure 12.5(b).

12.4 Watts–Strogatz small-world graphs

As can be expected, the average path length of a small-world graph scales linearly with the system size when $p = 0$ [4]:

$$l \approx n/2c. \tag{12.6}$$

As the rewiring probability p grows, the average path length decreases and approaches [4]

$$l \approx \frac{\ln(n)}{\ln(c)}. \tag{12.7}$$

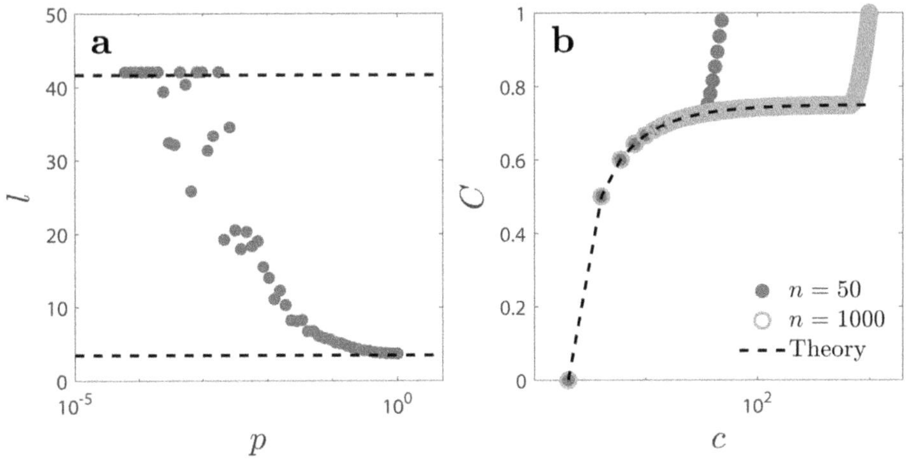

Figure 12.6. Average path length and clustering coefficient of Watts–Strogatz small-world graphs. (a) The average path length of a Watts–Strogatz small-world graph (blue dots) as a function of the rewiring probability p ($c = 6$, $n = 500$). The dashed lines represent the theoretical approximations for small and large values of p (equations (12.6) and (12.7), respectively). (b) Clustering coefficient (equation (12.3) of the Watts–Strogatz small-world graph as a function of c for $n = 50$ (blue dots) and for $n = 1000$ (orange circles). The results obtained using the theoretical formula (equation (12.8)) are added for comparison for very large values of n.

The clustering coefficient of a small-world graph also has different behaviors for different limiting cases. For the limit $p = 0$, there is an exact solution which gives the coefficient, as follows [4]:

$$C = \frac{3(c - 2)}{4(c - 1)}. \tag{12.8}$$

(Can you demonstrate this formula?) In the opposite limiting case, in which $p \to 1$, the graph practically becomes an Erdős–Rényi random graph with $p = c/(n - 1)$.

Interestingly, there is an intermediate range of p for which the Watts–Strogatz small-world graph has both a short path length and a large clustering coefficient, as observed in many real-world phenomena (see reference [4]).

Exercise 12.5. Average path length and clustering coefficient of Watts–Strogatz small-world graphs. Create a large Watts–Strogatz graph ($n > 200$, $c > 4$) for different values of p.

- **a.** Calculate the average path length of the graph for each p and plot your results. Verify the theoretical approximations for small and large values of p, as shown in figure 12.6(a). Repeat for different values of n and c.
- **b.** Calculate the clustering coefficient for $p = 0$ as a function of c and show that it matches the theoretical prediction, as shown in figure 12.6(b).

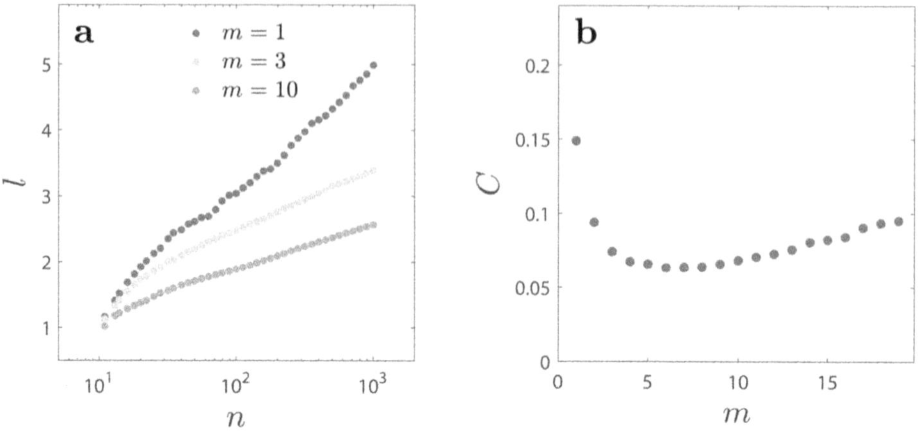

Figure 12.7. Average path length and clustering coefficient of Albert–Barabási preferential-growth graphs. (a) The average path length of an Albert–Barabási preferential-growth graph ($n_0 = 10$) as a function of the number of nodes for $m = 1$ (blue dots), $m = 3$ (yellow dots), and $m = 10$ (orange dots). (b) The clustering coefficient (equation (12.3)) of an Albert–Barabási preferential-growth graph (blue dots) as a function of the new connections per added node m. The other parameters are set to $n = 1000$ and $n_0 = 20$.

12.5 Albert–Barabási preferential-growth graphs

Unlike random networks such as Erdős–Rényi graphs, or partially random networks such as Watts–Strogatz graphs, the size (diameter) of the preferential-growth model always remains small, even if n is very large. Although there is no direct analytical solution for the clustering coefficient or the average path length of the preferential-growth graphs, there are theoretical approximations that indicate their behaviors, especially when n is very large.

Preferential-growth networks are known to be very small in diameter [5]. Unlike other random networks, the diameters of these graphs scale with $D \approx \ln[\ln(n)]$ for large values of n. In addition, the average path length scales according to $l \approx \ln(n)/\ln[\ln(n)]$. However, this limit is valid under the assumption that $\ln[\ln(n)] \gg 1$, which is numerically very challenging to reach ($\ln[\ln(n)] = 3.03$ for $n = 10^9$) [5]. Nevertheless, the dependence of l on n and m can still be studied, as demonstrated in Fig 12.7(a).

While the global clustering coefficient of a random network scales with $1/n$, the clustering coefficient of an Albert–Barabási preferential-growth model scales with $[\ln(n)]^2/n$ [6]. Another interesting study is the dependence of C on the number of nodes for m. It can be shown numerically that the dependence is not monotonic, i.e., there is an m value for every n that makes the graph minimally clustered, as shown in figure 12.7(b).

Exercise 12.6. Average path length and clustering coefficient of Albert–Barabási preferential-growth graphs. Create a large Albert–Barabási preferential-growth graph ($n = 1000$) for different values of m.

a. Calculate the average path length of the graph as a function of n for each m and plot your results. Compare your results with figure 12.7(a). *[Hint: You can calculate the* *average path length of the graph during the growth and record it as a function of n for computational efficiency.]*

b. Simulate Albert–Barabási graphs with $n_0 = 20$ and $m \in [1, 2, 3, \dots, 20]$ for $n = 1000$ and calculate the clustering coefficient of each graph. Compare your results with figure 12.7(b).

12.6 Further reading

For further information on network initialization and a more detailed introduction to the topic, see the reviews by Mark Newman [7], and Albert and Barabási [8]. Reference [7] is an excellent pedagogical review of the theoretical concepts, which includes real-world examples of networks.

Reference [3] is a great theoretical study that shows the dependence of the average path length and diameter of a random graph on the connection probability.

Reference [4] is a great source for theoretical derivations for small-world networks.

References [5, 6] are useful theoretical papers for a detailed study of the derivations of the average path length and the clustering coefficient of Albert–Barabási preferential-growth graphs.

12.7 Problems

Problem 12.1. Comparing network models. Generate Erdős–Rényi ($p = 0.02$), Watts–Strogatz ($c = 20, p = 0.01$), and Albert–Barabási ($m = 20$) graphs with $n = 1000$ nodes. The parameters are chosen in such a way as to roughly match the total number of edges. Calculate the average path length, the diameter, and the clustering coefficient of each graph. Compare them to each other and to the results of previous exercises. Comment on your results.

Problem 12.2. Analyzing a graph. Consider the graph in the file https://github.com/softmatterlab/SOCS/blob/main/Chapter_12_Network_Models/example_network.txt (which provides the adjacency matrix of the graph) on the GitHub repository. Analyze it and try to figure out the underlying network model.

a. Calculate its average path length and diameter.

b. Calculate its clustering coefficient.

c. Identify the underlying model for this network, i.e., find out whether this graph is more likely to be a Erdős–Rényi random graph, a Watts–Strogatz small-world graph, or an Albert–Barabási preferential-growth graph. *[Hint: You can use the insights gained from solving problem 12.7.1.]*

Problem 12.3. The files https://github.com/softmatterlab/SOCS/blob/main/Chapter_12_Network_Models/Network1.txt, https://github.com/softmatterlab/SOCS/blob/main/Chapter_12_Network_Models/Network2.txt, and https://github.com/softmatterlab/SOCS/blob/main/Chapter_12_Network_Models/Network3.txt respectively on the GitHub repository contain real-world data from three networks (in sparse format, i.e., as pairs of connected nodes). One is a social network of email exchanges at a Spanish university [9], one is the Western States power grid [4], and one is the (largest cluster of the) protein interaction network in yeast [10]. Your task is to use the tools you have learned about and constructed so far (e.g., degree distribution, clustering coefficient, diameter, and average path length) to identify which of these is which. What are the key properties of each network that sets it apart from the others?

Problem 12.4. Removing edges from a graph. Simulate an Albert–Barabási graph with a large number of nodes ($n > 500$) and a small number of connected edges per added node ($m < 5$).

a. Remove one edge randomly at a time and record the average path length of the network. Next, remove an edge connected to one of the nodes with the highest degree at the time and record the average path length as a function of the number of removed edges. Show that the latter approach results in a much greater increase of the average path length. Plot your results and compare them.

b. Change m and observe the ensuing differences.

c. Show that this difference in behavior is not nearly as pronounced in random graphs, such as Erdős–Rényi or Watts–Strogatz graphs.

Problem 12.5. Average path length for very large Albert–Barabási graphs. Generate an Albert–Barabási preferential-growth graph that has as many nodes as computationally practical ($n > 10^5$).

a. Measure the dependence of the average path length (as shown in figure 12.7(a)) as it becomes similar to the theoretical limiting behavior for $\ln[\ln(n)] \gg 1$ [5].

b. Measure the trend of the clustering coefficient for large n and show that it scales with $[\ln(n)]^2/n$ [6].

12.8 Challenges

Challenge 12.1. Alternative graph measures. There are plenty of additional graph measures that can be used to characterize graphs (e.g., efficiency, transitivity, rich clubs, communities, modularity). Check which measures are most useful in the characterization of the properties of networks with different underlying models. How effectively can they be used to identify the underlying model, given an adjacency matrix?

Challenge 12.2. From preferential growth to random graphs. How can you alter the preferential-growth model to make the degree distribution more like that of a random graph? Is it possible to introduce some mechanisms to limit the degree growth?

Challenge 12.3. Network activation. Given a graph, simulate its activation (using, e.g., Wilson-Cowan dynamics, a linear diffusion model, or the Fitzhugh–Nagumo model) at the level of each node. How can you infer the structure of the graph from the resulting time series? How does this depend on the network model used? *[Hint: Check the methods of reference [11] and references therein.]*

References

[1] Biggs N 1993 *Algebraic Graph Theory* (Cambridge University Press: Cambridge)

[2] Yang X, Lu S, Zhao W and Zhao Z 2019 Exploring the characteristics of an intra-urban bus service network: a case study of Shenzhen, China *ISPRS Int. J. Geo-Inf.* **8** 486

[3] Fronczak A, Fronczak P and Hołyst J A 2004 Average path length in random networks *Phys. Rev.* E **70** 056110
[4] Watts D J and Strogatz S H 1998 Collective dynamics of 'small-world' networks *Nature* **393** 440–2
[5] Cohen R and Havlin S 2003 Scale-free networks are ultrasmall *Phys. Rev. Lett.* **90** 058701
[6] Klemm K and Eguiluz V M 2002 Growing scale-free networks with small-world behavior *Phys. Rev.* E **65** 057102
[7] Newman M E J 2003 The structure and function of complex networks *SIAM Rev.* **45** 167–256
[8] Albert R and Barabási A-L 2002 Statistical mechanics of complex networks *Rev. Mod. Phys.* **74** 47–97
[9] Guimera R, Danon L, Diaz-Guilera A, Giralt F and Arenas A 2003 Self-similar community structure in a network of human interactions *Phys. Rev.* E **68** 065103
[10] Jeong H, Mason S P, Barabási A-L and Oltvai N Z 2001 Lethality and centrality in protein networks *Nature* **411** 41–2
[11] Mijalkov M, Pereira J B and Volpe G 2020 Delayed correlations improve the reconstruction of the brain connectome *PLoS One* **15** 1–22

Chapter 13

Evolutionary games

When interacting with other agents, it is crucial to identify the best possible strategy that leads to the best expected outcome. This is the goal of *game theory*. Examples include card games, political science, market pricing, and cooperation decisions [1]. One of the most famous game-theory examples is the *prisoner's dilemma* (figure 13.1). In its simplest version, two partners in crime have been apprehended and, now prisoners, are being interrogated separately. Let us call them prisoner A and prisoner B. If both remain silent, they will get just one year each. However, each of them can get a better deal by betraying their partner. For example, if A betrays B, A will get away free, but B will get fifteen years. If they both betray each other, they will both get five years. Therefore, it would be best for them as a group to remain silent and do their one year in prison. However, here is the catch: since the two prisoners are being interrogated independently, they do not have any means to know or control what their partner decides to do and, therefore, it is always convenient for them to betray each other: if their partner remain silent, they get away free; if their partner also betrays them, they get five years instead of fifteen. The dilemma arises because the choices that are best for individuals lead to a worse outcome for the group.

In most cases, agents need to make similar decisions time after time and can take advantage of their past experience [2–4]. Therefore, it is interesting to investigate how the behavior of individual agents changes during a series of similar interactions.

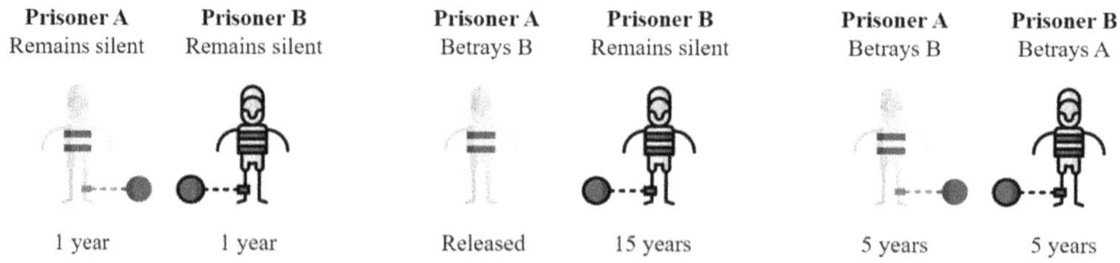

Figure 13.1. The prisoner's dilemma. The most famous game-theory example is the *prisoner's dilemma*. Although it is more profitable for prisoners as a group to cooperate (remain silent), it is always better from a single prisoner's perspective to defect (confess).

In this chapter, we first investigate the prisoner's dilemma between two players interacting for multiple rounds. We then consider a large group of players on a lattice playing the prisoner's dilemma with their nearest neighbors and updating their strategy according to their experience, and observe the emergence of *cooperative behavior*. Then, we consider how these interactions can be affected by *randomness*.

Example codes: Example Python scripts related to this chapter can be found at: https://github.com/softmatterlab/SOCS/tree/main/Chapter_13_Evolutionary_Games. Readers are welcome to participate in the discussions related to this chapter at: https://github.com/softmatterlab/SOCS/discussions/22.

13.1 The prisoner's dilemma

The original prisoner's dilemma is a two-player game that is often used as an example of the cooperation problem in game theory. As we have seen in the introduction, the dilemma arises from the fact that, from an individual's perspective, it is always more beneficial to defect against their partner in crime (betray them), even if the ideal solution for both as a group would be to cooperate (remain silent).

In its simplest form, the prisoner's dilemma has two partners in crime being interrogated separately, so that they cannot communicate. Both prisoners get an offer from the police (see figure 13.1): if they confess and witness against their partner, they will be punished by a mere T years (e.g., zero years) in prison, while their partner will be punished by S years in prison (e.g., fifteen years). However, if they both betray one another, both will be punished by P years in prison (e.g., five years). If they cooperate and both keep silent, they will only be punished by R years in prison (e.g., one year). Importantly, these punishments are arranged so that $T < R < P < S$. Depending on the specific conditions, T may be zero, or even a reward. To reduce the parameter space, we can assume that $T = 0$ and $P = 1$ with $R \in [0, 1]$ and $S \in [1, 2]$.

If the game is only played once, the best strategy is to defect. In fact, for each individual, it is clearly best to defect, as they will receive a smaller punishment regardless of what their partner does.

In a game with $N > 1$ rounds, the considerations change. In each round, both prisoners are going to either defect or cooperate. A strategy is to plan to cooperate for a certain number of rounds and defect afterwards. We denote this strategy by the number of intended cooperation rounds $n \in [0, N]$: cooperate until round n (included), or until the other player defects, and then defect for the rest of the game. This dictates that $n = 0$ is a strategy that always defects and $n = N$ is a strategy that always cooperates. A schematic representation of two players with strategies $n = 5$ and $m = 3$ is shown in figure 13.2.

The multiple-round prisoner's dilemma becomes interesting because, even though it is more beneficial for an individual prisoner to defect for a single round, it is also important to keep their partner cooperative for as many rounds as possible. Therefore, the best strategy is for a player to defect one round before their opponent

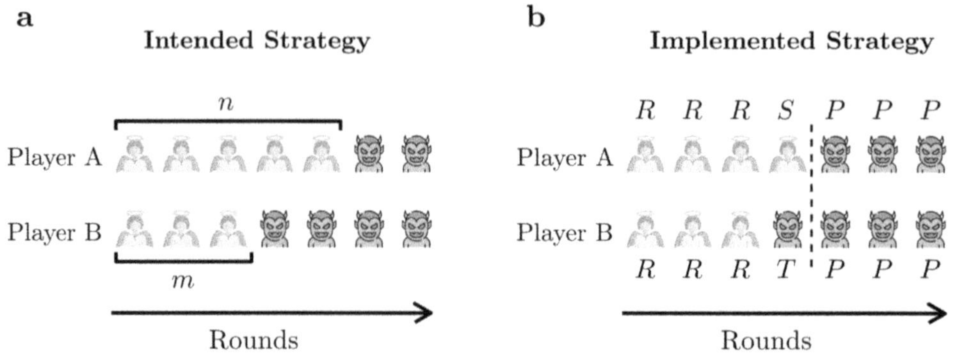

Figure 13.2. The prisoner's dilemma with multiple rounds. (a) Initially, players A and B have cooperative strategies (angel icon) for $n = 5$ and $m = 3$ rounds, respectively. (b) After player A is betrayed while cooperating (black dashed line), they start to defect (demon icon) and continue to defect for the rest of the game. At each round, if both players cooperate, they get R years in prison each; if only one cooperates, the cooperator gets S and the defector gets T; if they both defect, they both get P years ($T < R < P < S$).

is planning to. For example, let us consider a game with $N = 10$ rounds against a player with a strategy of $m = 6$. In this case, it is best to cooperate until the fifth round and defect in the sixth round (this is the $n = 5$ strategy), as shown in figure 13.3(a). A two-dimensional map of the resulting years in prison for prisoner A as a function of both n and m is shown in figure 13.3(b).

Exercise 13.1. Prisoner's dilemma with multiple rounds. Simulate the prisoner's dilemma for a number of rounds and calculate the accumulated years in prison. Initially, choose the parameters $N = 10$, $T = 0$, $R = 0.5$, $P = 1$, and $S = 1.5$.

- **a.** Fixing the opponent's strategy to m, show that the best strategy is $n = m - 1$, as shown in figure 13.3(a).
- **b.** Generate a 2D map of years in prison as a function of the player's strategy n and the opponent's strategy m, as shown in figure 13.3(b).
- **c.** While keeping $T = 0$ and $P = 1$, play with R and S to see how these parameters affect the results. Show that, as long as the essential conditions of the prisoner's dilemma ($T < R < P < S$) are satisfied, it is always best to use the strategy $m - 1$ against a player who uses strategy m.

13.2 Evolutionary games on a lattice

Although the underlying theory and the results are straightforward, even with multiple rounds, the prisoner's dilemma becomes complex when it is played many times with multiple players who have a memory of past games [2].

To study an ensemble of players that interact with each other, we consider a two-dimensional $L \times L$ array of players. At each time step, each player plays the prisoner's dilemma with their four closest neighbors, as shown in figure 13.4(a), and, if any of their neighbors achieves a better score, it updates its strategy, as shown in figure 13.4(b) [2, 3]. This evolutionary game can be simulated using the following steps:

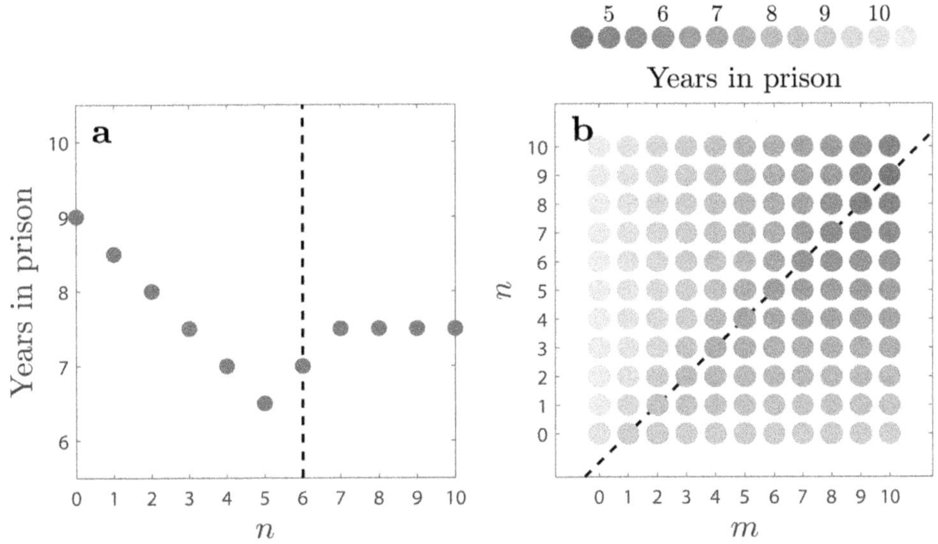

Figure 13.3. Optimal strategy for a multiple-round prisoner's dilemma game. (a) Years in prison as a function of strategy n playing against a player with strategy $m = 6$ (black dashed line). The best strategy is $n = 5$. This makes the opponent keep cooperating until a round before they defect. (b) 2D map of years of prison as a function of n and m. As shown here, the optimal strategy is always $n = m - 1$ (black dashed line). This is a game with $N = 10$ rounds with the parameters $T = 0$, $R = 0.5$, $P = 1$, and$S = 1.5$.

1. Initialize the $L \times L$ lattice with random strategies for each site, so that $n_{ij} \in [0, N]$ (see previous section for details about the strategy).
2. **Competition**: At each time step, each agent plays the prisoner's dilemma with its four nearest von Neumann neighbors (top, bottom, left, right), as shown in figure 13.4(a).
3. **Revision**: After the games have been simulated, each player updates its strategy to that with the best score (lowest punishment) amongst those of its

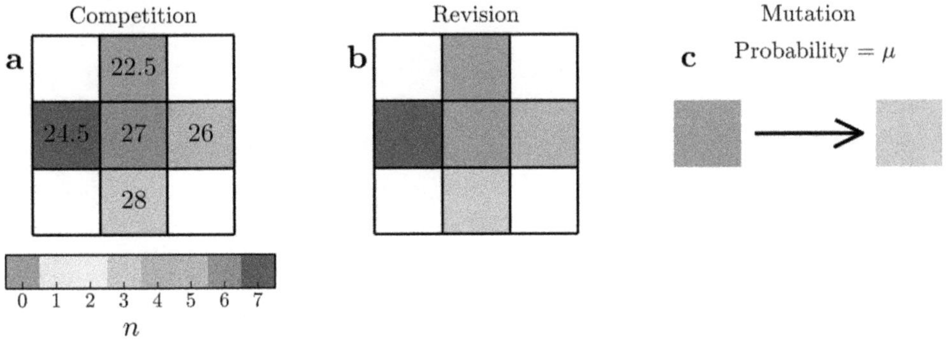

Figure 13.4. Prisoner's dilemma on a lattice. (a) At each time step, every player plays simple prisoner's dilemmas (N rounds) with its closest neighbors. Each player's score (lower is better) is shown by the numbers written inside each cell. (b) Then, every player adopts the strategy with the best score (lowest punishment, red square) amongst those played by its neighbors and itself. (c) Finally, some players mutate to a random strategy (probability μ).

neighbors and itself, as shown in figure 13.4(b). If multiple strategies tie, the choice is random between those.

4. **Mutation**: At the end of each time step, there is a small probability μ that each player will mutate its strategy to a random strategy, as shown in figure 13.4(c).

To understand the evolution of cooperation and defection, we start with players that can only use two strategies: The players can either cooperate in all rounds ($n = N$) or defect from the first round ($n = 0$). We first look at the case with no mutation ($\mu = 0$). For example, for $N = 7$, $R = 0.9$, and $S = 1.5$, if a single defector is placed in a uniform lattice of collaborators, it moves along the lattice and creates a pattern of defectors, as shown in figures 13.5(a)–(b). Different patterns can be obtained by initially inserting two (figures 13.5(c)–(d)), three (figures 13.5(e)–(f)), or four defectors (figures 13.5(g)–(h)). Yet different (and beautiful) patterns emerge when the parameters of the prisoner's dilemma or the number of neighbors are changed (e.g., using the Moore neighborhood with eight closest neighbors). It is even possible to obtain gliding and oscillating patterns, similar to those observed in the Game of Life (chapter 4) [5].

Exercise 13.2. Patterns in evolutionary games. Simulate the prisoner's dilemma on an $L \times L$ lattice. Allow only two strategies: always cooperate ($n = N$) and always defect ($n = 0$). Assume $\mu = 0$. Make sure that you choose a very small value of L at first to try your code and visualize its results. Then, increase L to a larger value (between 20 and 100). As usual, use $T = 0$ and $P = 1$. In addition, you can fix $S = 1.5$ and just play with the parameter R.

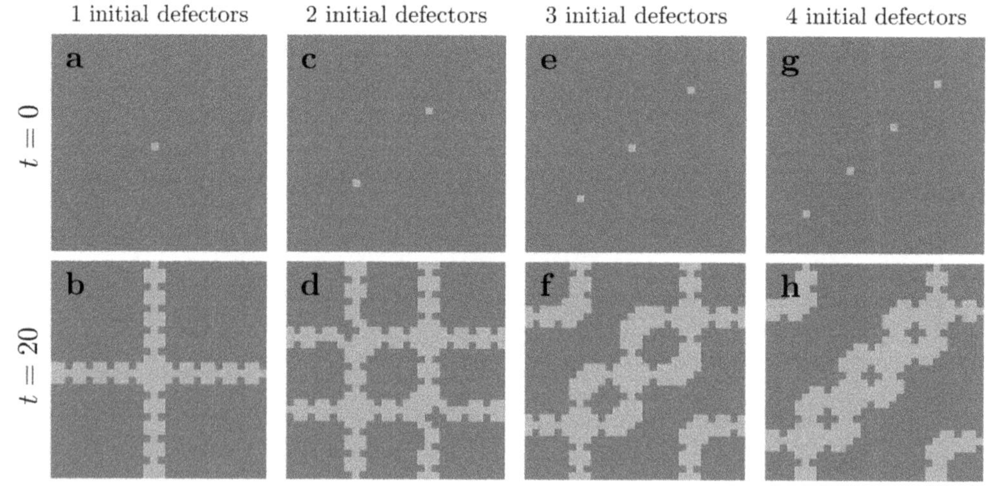

Figure 13.5. The spread of defectors amongst cooperators. (a) A single defector (orange) placed on a lattice of cooperators (blue) spreads, eventually forming the pattern of defectors in (b). Different patterns emerge for different initial numbers of defectors: (c)–(d) two initial defectors, (e)–(f) three initial defectors, and (g)–(h) four initial defectors. The parameters used for these simulations are $N = 7$, $R = 0.9$, $S = 1.5$, $L = 30$, and $\mu = 0$.

a. Initialize a lattice full of cooperators and place a single defector in the middle. Discover the range of R for which the defecting behavior only spreads in a line pattern in all directions, as shown in figure 13.5(b). What happens for other values of R?
b. Play with the number of initial defectors and simulate the system. Observe different pattern formations similar to those shown in figures 13.5(c)–(h).
c. What happens if you place a single cooperator in a lattice of defectors?
d. Find the ranges of R for which a cluster of cooperators in a background of defectors vanishes, stays stable, or spreads.

With a non-zero mutation rate ($\mu \neq 0$), three regimes emerge, depending on the parameter R (or S, whichever we decide to change). The regime in which cooperation dominates (as shown in figure 13.6(a)), the regime in which cooperators and defectors coexist (as shown in figure 13.6(b)), and the regime in which defection dominates (as shown in figure 13.6(c)).

Exercise 13.3. Defectors vs cooperators. Simulate the prisoner's dilemma on an $L \times L$ lattice using only two strategies: always cooperate ($n = N$) and always defect ($n = 0$). Use a small but non-zero mutation rate, such as $\mu = 0.01$. As usual, use $T = 0$ and $P = 1$. In addition, you can fix $S = 1.5$ and just play with the parameter R.
a. With $R = 0.82$, show that cooperation dominates, as shown in figure 13.6(a).
b. With $R = 0.84$, show that cooperation coexists with defection, as shown in figure 13.6(b).
c. With $R = 0.86$, show that cooperation vanishes (figure 13.6(c)).
d. Repeat your simulations for a range of R to show that this behavior is critical. Determine the critical values of R that lead to each regime.
e. Repeat the same analysis, fixing the value of R and varying S.

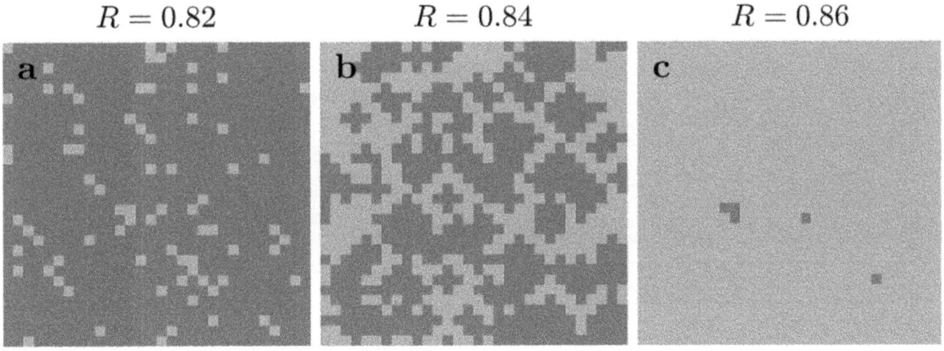

Figure 13.6. Defectors vs cooperators. (a) When the punishment for cooperating prisoners is low ($R = 0.82$), the cooperation strategy (blue) dominates, although small clusters of defectors (orange) can still exist. (b) When the punishment for cooperating prisoners is higher ($R = 0.84$), cooperation (blue) coexists with defection (orange). (c) When the punishment for cooperating prisoners is even higher ($R = 0.86$), the cooperation strategy (blue) disappears, eventually leading to the total domination of defectors (orange). Snapshots taken after 100 time steps. Parameters: $N = 7$, $S = 1.5$, $L = 30$, and $\mu = 0.01$.

13.3 Multiple strategies

So far, we have considered evolutionary games with only two strategies: always cooperate ($n = N$) and always defect ($n = 0$). However, one can also consider evolutionary games in which a player cooperates for an arbitrary number of rounds and then defects. In the case in which an agent plays against a single agent, as shown in figure 13.3, the outcome is straightforward. However, when the players play this game on a lattice with multiple agents that have different strategies, the simulations become more complex and interesting. In this section, we will generalize our simulations to multiple strategies with $n \in [0, N]$.

Under these conditions, small clusters of players with the same strategy may emerge, propagate, and disappear depending on the choice of parameters. With $S = 1.5$, cooperative strategies prevail for small R (R(0.50), while all strategies coexist together for intermediate values of R (0.50(R)0.75), and defection strategies dominate for large R (R(0.75). Figure 13.7 shows the outcome for $R = 0.72$; at this value, different strategies appear, grow, and propagate. Importantly, there are large fluctuations, as shown in figure 13.7(b).

Exercise 13.4. Evolutionary games on a lattice with multiple strategies. Simulate a seven-round prisoner's dilemma on an $L \times L$ lattice. Allow all strategies ($0 \leqslant n \leqslant 7$). Use a small but non-zero mutation rate, such as $\mu = 0.01$. As usual, use $T = 0$ and $P = 1$. Depending on the parameters R and S, multiple regimes can be observed.

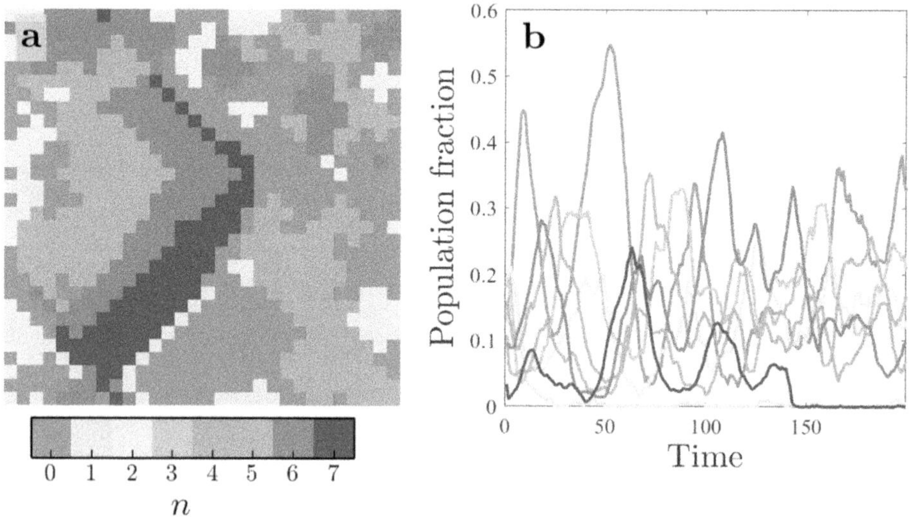

Figure 13.7. Evolutionary games on a lattice with multiple strategies. (a) Snapshot of the strategies obtained after 200 steps of the simulation. Different strategies coexist, often as clusters with diagonal borders (matching the neighborhood borders). (b) The population fraction with different strategies during the simulation. Different strategies continuously emerge, grow, and vanish. In general, a large population of defectors (red tones) favors the growth of cooperative behavior (blue tones), and vice versa. Parameters: $N = 7$, $R = 0.72$, $S = 1.5$, $L = 30$, and $\mu = 0.01$.

a. Fix $S = 1.5$ and play only with parameter R. Observe what happens within the lattice as time evolves. Show that different strategies may emerge and propagate, as shown in figure 13.7(a).
b. Show that there are three main regimes that are similar to those identified in exercise 13.3. However, this time, the regime in which different strategies coexist can have different dominant populations, depending on the value of R. Discover the parameters and the population distribution.
c. Discuss the results of your simulation. What do these numerical outcomes tell us about the evolution of cooperation? Which strategies are *stable evolutionary strategies*?

Exercise 13.5. Two-dimensional phase map of evolutionary games. For the case of multiple allowed strategies, the dynamics does not only depend on one parameter, or a simple mathematical relation between the two parameters R and S. Instead, the output behavior is rather complex and depends on both parameters.
a. Simulate the evolutionary games with different values of R and S. For every simulation, record at least 500 steps and omit the first 100 steps in the analysis. Calculate the variance of the population for each strategy σ_n^2.
b. $\sum_n \sigma_n^2$ can be used to determine whether there is active competition between strategies. Determine a reasonable threshold for this parameter and create a phase diagram for the output behavior as a function of R and S.

13.4 Further readings

Reference [1] is an excellent book for reviewing different models of cooperative game theory; it also includes methodological principles, solution concepts, and model analyses.

Reference [4] is a great book for theoretical information on evolutionary games; it also deals with the implications and connections between evolutionary biology and cooperative strategies in economics.

Reference [2] is the pioneering work on evolutionary game theory with players that have a memory of past games. Reference [3] is an excellent source for the evolution of cooperation with numerical simulations.

Reference [5] demonstrates that the rules in evolutionary games can be tuned so that the output behaviors resemble those encountered in the Game of Life.

Reference [6] is a recent study that demonstrates the emergence of exploitative behavior when players use reinforcement learning to develop their probabilistic strategies. Interestingly, it is shown that this occurs even when the rules of the game and the strategy dynamics are symmetric.

13.5 Problems

Problem 13.1. Pattern emergence. Consider different neighborhoods for evolutionary games with no mutation. What patterns emerge? What are the differences? *[Hint: For each neighborhood, the parameters for the formation of patterns may vary, so you may need to find the right parameters first.]*

a. Repeat exercise 13.2 with Moore neighbors (i.e., in each step, the player plays with their eight closest neighbors).
b. Implement competition with von Neumann neighbors and revision with Moore neighbors.
c. Implement competition with Moore neighbors and revision with von Neumann neighbors.
d. Increase the interaction range to two and three lattice sites in each direction.

Problem 13.2. Probabilistic strategies. Assume that the strategy n of each agent in an evolutionary game lattice is probabilistic, i.e., an agent with strategy n may apply the strategy $n \pm 1$ with probability ϵ.
a. Simulate a lattice of $L \times L$ agents playing the prisoner's dilemma. Determine the values of R for which different strategies become dominant as a function of ϵ.
b. Suppose that the agents have a probability p_n of applying each strategy n. Simulate this system and observe how the population evolves.

Problem 13.3. Evolutionary games on a graph. Consider agents playing the prisoner's dilemma on a graph, i.e., with agents that they are connected to by an edge.
a. Use an Erdős–Rényi random graph (see exercise 12.1) with $N = 100$ nodes and a connection probability of $p = 0.1$. Observe the dynamics.
b. Now use a Watts–Strogatz small-world graph (see exercise 12.2) of comparable degree. What are the differences?
c. Finally, use an Albert–Barabási preferential-growth graph (see exercise 12.3). Calculate the average strategy number n as a function of the degree of the nodes.
d. Try to update the connections at the end of each step, instead of updating the strategies of each agent: each agent randomly rewires one of its connections with an agent that betrayed it (before it could betray them). Calculate the evolution of the network's clustering coefficient, average path length, diameter, and the degree distribution as a function of the time steps.

13.6 Challenges

Challenge 13.1. Neural networks for the optimum strategy. Implement a neural network that takes the history of an agent's strategies, its neighbors' strategies, and its score as an input in order to predict the best strategy to maximize its own benefit.
a. Train the network using a lattice of regular players (figure 13.4) and discuss the ensuing behavior.
b. Create a lattice in which each agent updates itself according to the strategy suggested by the neural network. Discuss the ensuing dynamics.

Challenge 13.2. Gliders and rotators. Under certain initial conditions and with an appropriate choice of neighborhood, a cluster of cooperators may glide or rotate, like the structures in the Game of Life. Investigate these formations and find out which structures and parameters result in gliding and rotation. Can you reproduce other aspects of the Game of Life? *[Hint: Check reference [5].]*

Challenge 13.3. Emergence of exploitative relationships. Consider an evolutionary game of the prisoner's dilemma in which players update their strategy using reinforcement learning. Show that, for certain parameters, an exploitative behavior emerges, i.e., some players accept their antagonist's defection to optimize their strategy. *[Hint: Check reference [6].]*

References

[1] Myerson R B 2013 *Game Theory* (Cambridge, MA: Harvard University Press)

[2] Axelrod R and Hamilton W D 1981 The evolution of cooperation *Science* **211** 1390–6

[3] Nowak M A and May R M 2020 Evolutionary games and spatial chaos *Nature* **359** 826–9

[4] Weibull J W 1997 *Evolutionary Game Theory* (Cambridge, MA: MIT Press)

[5] Nowak M A and May R M 1993 The spatial dilemmas of evolution *Int. J. Bifurc. Chaos* **3** 35–78

[6] Fujimoto Y and Kaneko K 2019 Emergence of exploitation as symmetry breaking in iterated prisoner's dilemma *Phys. Rev. Res.* **1** 033077

Chapter 14

Ecosystems

Ecosystems are where life perpetuates itself, thanks to the complex dynamics resulting from the interaction between inanimate elements (e.g., the ground, water, the atmosphere, weather conditions, orographic morphology) and animated ones (e.g., animals, plants, fungi, bacteria). Each ecosystem is characterized by a delicate and complex dynamical equilibrium between its components. For example, let us consider a simple rural ecosystem with plants, insects, and birds: The birds eat the insects, which in turn eat the plants. If the birds are heavily hunted, the number of insects will grow, in turn causing serious damage to the plants. Also the introduction of foreign species into an ecosystem can have unexpected and often damning consequences. For example, rats were inadvertently introduced into remote islands by exploratory ships, causing the extinction of many endemic species of birds, mammals, reptiles, invertebrates, and plants. As another example, the rosy wolfsnail (*Euglandina rosea*) was purposefully introduced in the South Pacific islands as a biocontrol agent for the giant African land snail (*Lissachatina fulica*), but it found it

Figure 14.1. A prey–predator interaction. To sustain themselves and their offspring, predators hunt their prey: this robin dug a worm out of the grass for its next meal. *Source: Unsplash: Mathew Schwartz.* https://unsplash.com/photos/rQ2_FONfh04/photo

doi:10.1088/978-0-7503-3843-1ch14

easier to hunt other endemic snail species, which have now become extinct [1, 2]. It is therefore of the utmost importance to understand the dynamics of ecosystems.

In this chapter, we start by analyzing a classical model for an important interaction in ecosystems: the *prey–predator interaction* (figure 14.1). The first prey–predator model was introduced by Vito Volterra in 1926. This model consists of a system of two first-order differential equations. We use this as a toy model to exemplify the study of a *dynamical system*, connecting the analytical properties of the model, such as the *equilibrium points* and their *stability*, its linearized version near the equilibrium points, the presence of *invariants*, and the solution found by a finite-difference numerical integration scheme. We emphasize the thoughtful use of the numerical tools by exploring three different techniques: the explicit, implicit, and symplectic Euler integration schemes. We then review a few other models of important interactions found in ecosystems: the *logistic model*, which deals with the interaction between a single species and its environment, a few models for *mutualism*, and a model for *competition*. In all these cases, we benefit from the prototype analysis conducted for the case of the prey–predator system.

Example codes: Example Python scripts related to this chapter can be found at: https://github.com/softmatterlab/SOCS/tree/main/Chapter_14_Ecosystems. Readers are welcome to participate in the discussions related to this chapter at: https://github.com/softmatterlab/SOCS/discussions/23.

14.1 Lotka–Volterra model

A mathematical model of a prey–predator system was first introduced by Vito Volterra, while trying to explain the unusual data for fish catches in the Adriatic Sea during the years of World War I [3]. As he was inspired by the Lotka theory of autocatalytic chemical reactions [4], the model takes the name Lotka–Volterra. The equations describing the interaction between the prey x and the predators y are the following:

$$\begin{cases} \dfrac{\mathrm{d}x}{\mathrm{d}t} = \alpha x - \beta xy \\ \dfrac{\mathrm{d}y}{\mathrm{d}t} = -\gamma y + \delta xy \end{cases} \tag{14.1}$$

where α is the exponential growth rate of the prey, β is the decline rate in the prey population due to predation, γ is the exponential decline in population of the predators (predators are thought to go extinct if no prey are available), and δ is the growth rate in the predator population resulting from predation.

If we have a certain number of prey x and predators y at a given instant t, system (14.1) tells us the instantaneous variation of each population $\left(\frac{\mathrm{d}x}{\mathrm{d}t}, \frac{\mathrm{d}y}{\mathrm{d}t}\right)$. For each $x \geqslant 0$ and $y \geqslant 0$, we can then represent the *direction* in which the system population

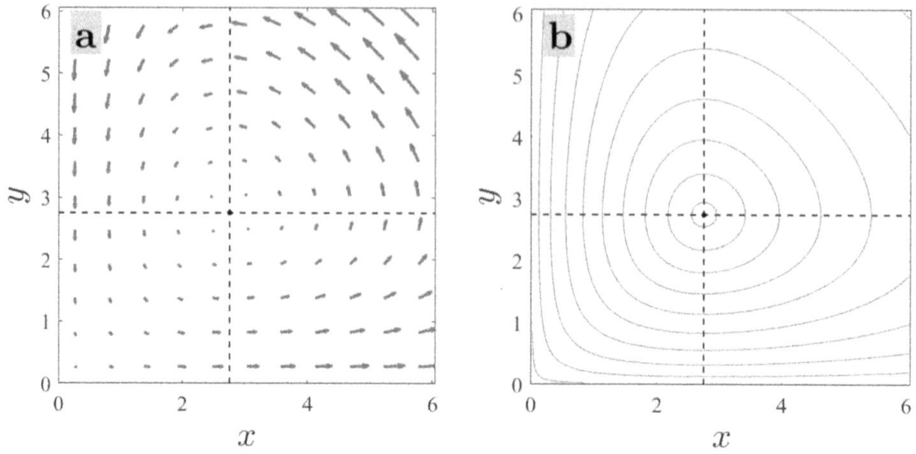

Figure 14.2. Vector field and invariant of a Lotka–Volterra system. (a) Vector field of a Lotka–Volterra system. The equilibrium points are (0, 0) (unstable equilibrium) and $(\frac{\gamma}{\delta}, \frac{\alpha}{\beta})$ (stable equilibrium, black dot). The dashed black lines indicate where $\frac{dx}{dt} = 0$ and $\frac{dy}{dt} = 0$. (b) Level curves of the invariant $I(x, y)$ (equation (14.3)). All level curves are closed around the stable equilibrium point, indicating that the trajectories must be periodic.

evolves $(v_x, v_y) = \left(\frac{dx}{dt}, \frac{dy}{dt}\right)$ (figure 14.2(a)). This is called the *vector field* associated with the system (14.1).

The system (14.1) has two *equilibrium points* (i.e., two values of the initial populations x_0 and y_0 that lead to constant solutions over time: $x(t) = x_0$ and $y(t) = y_0$):

$$\begin{cases} x_{eq} = \frac{\gamma}{\delta} \\ y_{eq} = \frac{\alpha}{\beta} \end{cases} \text{and} \begin{cases} x_0 = 0 \\ y_0 = 0 \end{cases}. \tag{14.2}$$

It is easy to verify that with these initial conditions, system (14.1) has a constant solution, because the time derivatives of $x(t)$ and $y(t)$ vanish. If, instead, we start from a generic condition $x(0) = x_0$, $y(0) = y_0$, the derivatives $\frac{dx}{dt}$ and $\frac{dy}{dt}$ are not simultaneously zero and at least one of the two populations varies.

The quantity $I(x, y)$, defined as

$$I(x, y) = \delta x - \gamma \log x + \beta y - \alpha \log y \tag{14.3}$$

is an *invariant*, i.e., it is a conserved quantity for each trajectory in phase space. As a consequence of this, each trajectory lies on a level curve of $I(x, y)$. These curves must be closed (figure 14.2(b)) and, therefore, the trajectories are periodic.

Exercise 14.1. Invariant. Show that $I(x, y)$ (equation (14.3)) is a an invariant of the Lotka–Volterra system (14.1). *[Hint: From system (14.1), obtain a relation for* $\frac{dx}{dt}/\frac{dy}{dt}$ *in terms of* xy, α, β, γ, *and* δ. *Separate the variables and integrate in time.]*

The closer we get to the stable equilibrium point $(x_{eq}, y_{eq}) = (\frac{\gamma}{\delta}, \frac{\alpha}{\beta})$, the more the trajectories resemble a circle in phase space. Close to this equilibrium point, we can linearize the Lotka–Volterra equations. To do this, substitute $x(t) = \frac{\gamma}{\delta} + u(t)$ and $y(t) = \frac{\alpha}{\beta} + v(t)$ into equation (14.1) to obtain:

$$\begin{cases} \frac{du}{dt} = -\frac{\beta\gamma}{\delta}v - \beta uv \\ \frac{dv}{dt} = \frac{\alpha\delta}{\beta}u + \delta uv \end{cases} \tag{14.4}$$

where the facts that $\frac{du}{dt} = \frac{dx}{dt}$ and $\frac{dv}{dt} = \frac{dy}{dt}$ have been used. The system (14.4) is still nonlinear because of the presence of the uv terms However, as $u(t) \approx 0$ and $v(t) \approx 0$ (why?), we can linearize the system by dropping the uv terms, obtaining

$$\begin{cases} \frac{du}{dt} = -\frac{\beta\gamma}{\delta}v \\ \frac{dv}{dt} = \frac{\alpha\delta}{\beta}u \end{cases} \tag{14.5}$$

which has immediate similarities with the simple harmonic oscillator equation. We can see this by taking the time derivative of the first equation of (14.5), $\frac{d^2u}{dt^2} = -\frac{\beta\gamma}{\delta}\frac{dv}{dt}$, and substituting the second equation of (14.5) into its second term, which leads to $\frac{d^2u}{dt^2} = -\alpha\gamma\, u$, which is the equation for a harmonic oscillator with an angular frequency $\omega = \sqrt{\alpha\gamma}$. For initial conditions of $u(0) = u_0$ and $v(0) = v_0$, the solution of the linearized system (14.5) is

$$\begin{cases} u(t) = A\cos(\omega t + \phi) \\ v(t) = \frac{\delta}{\beta\gamma}A\omega\sin\left(\omega t + \phi\right) \end{cases} \tag{14.6}$$

where A and ϕ depend on the initial conditions, as follows:

$$A = \sqrt{u_0^2 + \left(\frac{\beta\gamma}{\delta\omega}v_0\right)^2} \text{ and } \cos\phi = \frac{u_0}{A}, \quad \sin\phi = \frac{1}{A\omega}\frac{\beta\gamma}{\delta}v_0.$$

Therefore, the prey and predators oscillate around their respective equilibrium values. The solutions of the linearized system are all periodic, with the same period $T_0 = \frac{2\pi}{\omega}$, which is independent of their initial conditions. (T_0 is also the limit value of the period of the solutions of the initial system (14.1), when the initial condition

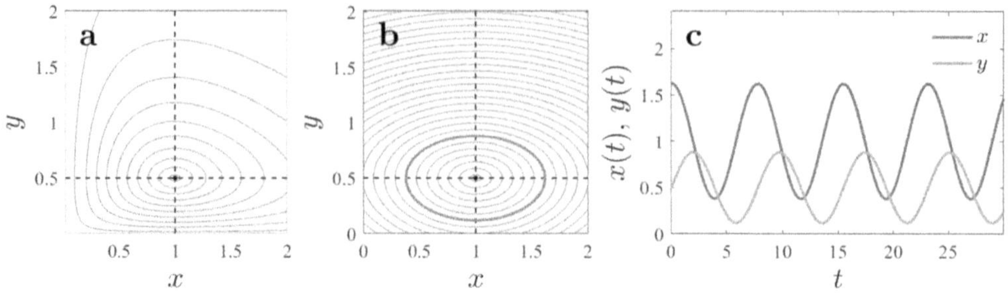

Figure 14.3. Linearized Lotka–Volterra model. (a) Invariant level curves of a Lotka–Volterra system (equations (14.1), $\alpha = 2/3$, $\beta = 4/3$, $\gamma = 1$, and $\delta = 1$) and (b) of the corresponding linearized system (equations (14.5)). The green line highlights the level curve for one solution. (c) Analytical solutions for $x(t)$ and $y(t)$ obtained using the solutions $u(t)$ and $v(t)$ of the linearized systems (equations (14.6)) for initial conditions $x_0 = x_{eq} + u_0 = 1.62$ and $y_0 = y_{eq} + v_0 = 0.47$.

$(x_0, y_0) \to (x_{eq}, y_{eq})$). The prey peak anticipates the predator peak by exactly a quarter of a period (i.e., it is dephased by exactly $\frac{\pi}{2}$).

Exercise 14.2. Linearized Lotka–Volterra model. Plot the solutions of the linearized Lotka–Volterra model (equations (14.6)) for various initial conditions (x_0, y_0) near the equilibrium point. Compare your results with figure 14.3.

Explicit Euler integration scheme. We now proceed to the numerical integration of the population dynamics described by equations (14.1), using finite-difference equations. First, we employ the *explicit Euler integration scheme* (analogously to what we did in chapter 1 on molecular dynamics). Given the population numbers at time step n, the population numbers at time step $n + 1$ are

$$\begin{cases} x_{n+1} = x_n + \left(\alpha\, x_n - \beta\, x_n\, y_n\right) \Delta t \\ y_{n+1} = y_n + \left(\delta\, x_n\, y_n - \gamma\, y_n\right) \Delta t \end{cases} \tag{14.7}$$

where Δt is the length of the time step.

Exercise 14.3. Explicit Euler integration of the Lotka–Volterra model.

a. Write some code implementing the explicit Euler integration scheme described in equations (14.7). Take, for instance, $\alpha = 2/3$, $\beta = 4/3$, $\gamma = 1$, $\delta = 1$, $x_0 = 1$, and $y_0 = 1$. Show that the numerical solution is very similar to figure 14.4(a).

b. Calculate and plot the value of the invariant $I(x, y)$ (equation (14.3)) along a trajectory. Is it constant in time? Compare it with figure 14.4(b).

c. Plot your solution in phase space. Is it periodic? What happens to this trajectory in the long run? Compare it with figure 14.4(c).

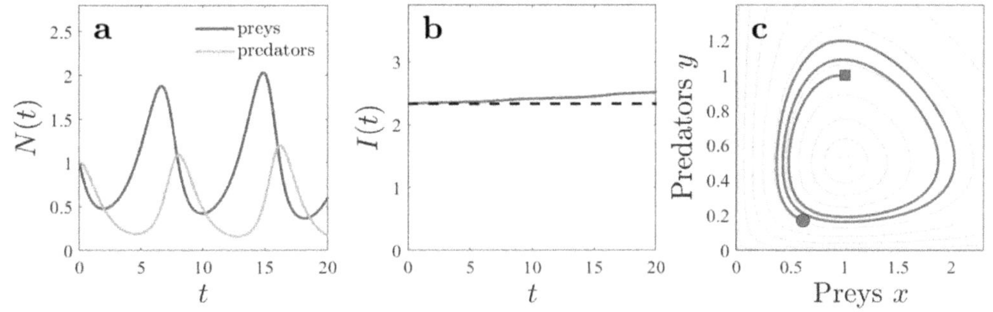

Figure 14.4. Explicit Euler integration of the Lotka–Volterra model. (a) Numerical solutions $x(t)$ and $y(t)$ of the Lotka–Volterra model using the explicit Euler integration scheme (equations (14.7)). (b) $I(x, y)$ versus time, featuring a drift over time. (c) Trajectory in phase space; the initial point is marked with a square and the final point is marked with a circle. It spirals away from the equilibrium point.

Similarly to what we saw in chapter 1 when using the Euler scheme to simulate molecular dynamics, we find that the Euler scheme is also biased when simulating the Lotka–Volterra model. The numerical solution given by this integration method is not periodic (figure 14.4(a)). Furthermore, the invariant $I(x, y)$ calculated along the trajectory is not constant in time, but features an increasing trend (figure 14.4(b)), and the trajectory in phase space is not closed (figure 14.4(c)). Therefore, the explicit Euler integration scheme has flaws that might undermine the stability of the numerical simulation and the plausibility of its results.

Implicit Euler integration scheme. A way to overcome the problems of the explicit Euler integration scheme is to use an *implicit Euler integration scheme*. Given a system

$$\begin{cases} \dfrac{dx}{dt} &= f(x, y) \\ \dfrac{dy}{dt} &= g(x, y) \end{cases} \tag{14.8}$$

an implicit scheme for the solution of the associated finite-difference equation is defined as follows:

$$\begin{cases} x_{n+1} &= x_n + f(x_{n+1}, y_{n+1})\, \Delta t \\ y_{n+1} &= y_n + g(x_{n+1}, y_{n+1})\, \Delta t \end{cases} \tag{14.9}$$

where x_{n+1} and y_{n+1} are implicitly defined and usually determined by solving a nonlinear equation. (By contrast, an *explicit* scheme would use x_n and y_n on the right-hand side.) Therefore, the implicit Euler integration scheme for the Lotka–Volterra model (equations (14.1)) is

$$\begin{cases} x_{n+1} &= x_n + \left(\alpha x_{n+1} - \beta x_{n+1} y_{n+1}\right) \Delta t \\ y_{n+1} &= y_n + \left(\delta x_{n+1} y_{n+1} - \gamma y_{n+1}\right) \Delta t \end{cases} \tag{14.10}$$

Exercise 14.4. Implicit Euler integration of the Lotka–Volterra model.

a. Solve the nonlinear system (14.10) to find the unknown x_{n+1}, y_{n+1} as a function of x_n, y_n, Δt, α, β, γ, and δ. *[Hint: First, express x_{n+1} and y_{n+1} as functions of the other terms, i.e., $x_{n+1} = x_n(1 - \Delta t(\alpha - \beta y_{n+1}))^{-1}$ and $y_{n+1} = y_n(1 - \Delta t(\delta x_{n+1} - \gamma))^{-1}$. Then, substitute the expression for x_{n+1} into that for y_{n+1}, and* vice versa. *With some further algebraic steps, you will obtain*

$$(1 - \alpha\Delta t)\delta\Delta t x_{n+1}^2 - [(1 - \alpha\Delta t)\cdot(1 + \gamma\Delta t) + \delta\Delta t\, x_n + \beta\Delta t y_n]x_{n+1} + (1 + \gamma\Delta t)x_n = 0$$

and

$$\left(1 + \gamma\Delta t\right)\beta\Delta t y_{n+1}^2 + [(1 - \alpha\Delta t)\cdot(1 + \gamma\Delta t) - \beta\Delta t y_n - \delta\Delta t x_n]y_{n+1} - \left(1 - \alpha\Delta t\right)y_n = 0,$$

which are two second-degree algebraic equations for x_{n+1} and y_{n+1}. Solve them in the standard way and choose the solution that is closest to (x_n, y_n).]

b. Using the formulas obtained in part (a), implement a function that calculates $(\tilde{x}_{n+1}, \tilde{y}_{n+1})$ given (x_n, y_n) and use it to solve the Lotka–Volterra model using the implicit Euler integration scheme (equations (14.10)), i.e., use the estimate $\tilde{x}_{n+1}$, $\tilde{y}_{n+1}$ to calculate

$$\begin{cases} x_{n+1} = x_n + \left(\alpha\tilde{x}_{n+1} - \beta\tilde{x}_{n+1}\tilde{y}_{n+1}\right)\Delta t \\ y_{n+1} = y_n + \left(\delta\tilde{x}_{n+1}\tilde{y}_{n+1} - \gamma\tilde{y}_{n+1}\right)\Delta t \end{cases}.$$

c. Plot the numerical solution found. Compare it with figure 14.5(a).

d. Calculate the value of the invariant $I(x, y)$. Is it constant in time? Compare it with figure 14.5(b).

e. Plot the solution in phase space. Is it a closed orbit? Compare it with figure 14.5(c).

The solution of the implicit Euler integration scheme is also non-periodic (figure 14.5(a)); its anticipated invariant $I(x, y)$ is not constant but steadily decreases

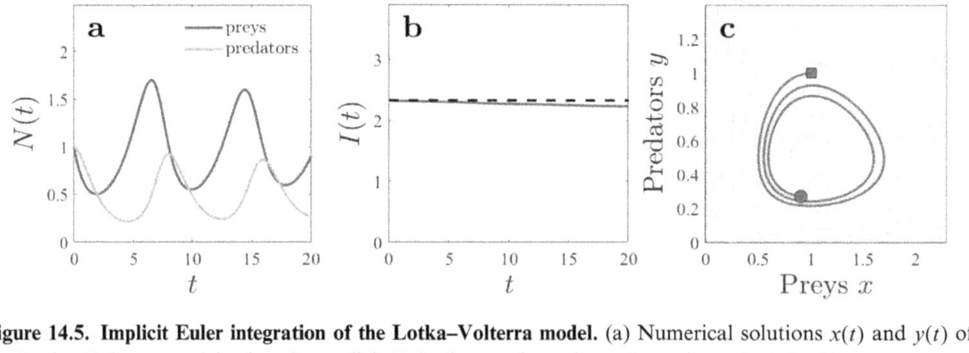

Figure 14.5. Implicit Euler integration of the Lotka–Volterra model. (a) Numerical solutions $x(t)$ and $y(t)$ of the Lotka–Volterra model using the explicit Euler integration scheme (equations (14.10)). (b) $I(x, y)$ versus time, featuring a drift over time. (c) Trajectory in phase space; the initial point is marked by a square and the final point is marked by a circle. It gradually converges toward the equilibrium point.

(figure 14.5(b)), and the trajectory spirals in phase space (figure 14.5(c)). However, in contrast to the explicit Euler integration scheme, the implicit scheme is more stable: if there is an equilibrium point, the trajectory evolves and approaches the equilibrium point. For this reason, implicit methods are very good for finding equilibrium points and attractors, even though they still do not provide reliable numerical integration.

Symplectic Euler integration scheme. If an explicit scheme pushes the simulated trajectory inexorably away from the equilibrium point, and an explicit integration pulls it toward the equilibrium point, can we somehow combine them to obtain the correct solution?

One way to do this is to use a *symplectic Euler integration scheme*, which treats one variable as implicit and the other as explicit. Thus, we have the two following possibilities:

$$\begin{cases} x_{n+1} = x_n + \left(\alpha x_n - \beta x_n y_{n+1}\right)\Delta t \\ y_{n+1} = y_n + \left(\delta x_n y_{n+1} - \gamma y_n\right)\Delta t \end{cases} \quad \text{with } x \text{ explicit and } y \text{ implicit,} \tag{14.11}$$

or

$$\begin{cases} x_{n+1} = x_n + \left(\alpha x_{n+1} - \beta x_{n+1} y_n\right)\Delta t \\ y_{n+1} = y_n + \left(\delta x_{n+1} y_n - \gamma y_{n+1}\right)\Delta t \end{cases} \quad \text{with } x \text{ implicit and } y \text{ explicit.} \tag{14.12}$$

Exercise 14.5. Symplectic Euler integration of the Lotka–Volterra model.

a. Derive the algebraic expressions for the estimate of x_{n+1} and y_{n+1} for systems (14.11) and (14.12). Proceed in a similar way to that used for exercise 14.4 (this time, the algebra is much simpler). *[Hint: Show that, for equations (14.11), the estimate for* $(\tilde{x}_{n+1}, \tilde{y}_{n+1})$ *is*

$$\begin{cases} \tilde{x}_{n+1} = x_n\left(1 + \Delta t\left(\alpha - \dfrac{\beta\, y_n}{1 - \Delta t(\delta\, x_n - \gamma)}\right)\right) \\ \tilde{y}_{n+1} = \dfrac{y_n}{1 - \Delta t(\delta\, x_n - \gamma)} \end{cases} \quad x \text{ explicit, } y \text{ implicit,} \tag{14.13}$$

and, for equations (14.12), the estimate for $(\tilde{x}_{n+1}, \tilde{y}_{n+1})$ *is*

$$\begin{cases} \tilde{x}_{n+1} = \dfrac{x_n}{1 - \Delta t(\alpha - \beta\, y_n)} \\ \tilde{y}_{n+1} = y_n\left(1 + \Delta t\left(\dfrac{\delta\, x_n}{1 - \Delta t(\alpha - \beta\, y_n)} - \gamma\right)\right) \end{cases} \quad x \text{ implicit, } y \text{ explicit. }] \tag{14.14}$$

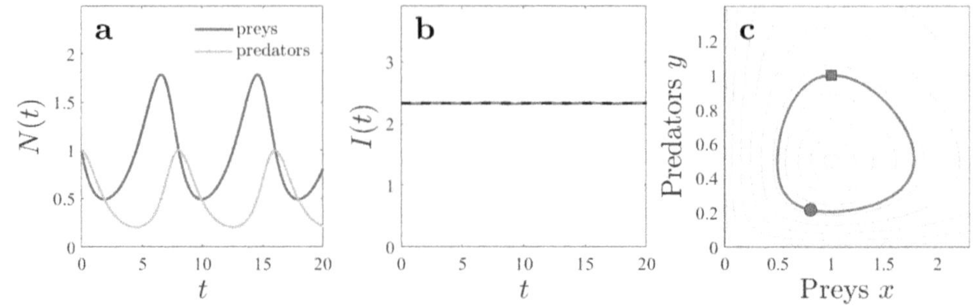

Figure 14.6. Symplectic Euler integration of the Lotka–Volterra model. Numerical solutions $x(t)$ and $y(t)$ of the Lotka–Volterra model obtained using the symplectic Euler integration scheme (equations (14.12), x implicit and y explicit). (b) $I(x, y)$ versus time, which does not have a drift (even though it still slightly fluctuates around its average value). (c) Trajectory in phase space; the initial point is marked by a square and the final point is marked by a circle. The trajectory is periodic. The other symplectic Euler integration scheme (equations (14.11), x explicit and y implicit) gives similarly accurate results.

b. Using the formulas obtained in the previous point, implement some functions to find $(\tilde{x}_{n+1}, \tilde{y}_{n+1})$. Use them to solve the Lotka–Volterra model using the respective symplectic Euler integration schemes (equations (14.13) and (14.14)). Plot the numerical solutions and compare them with figure 14.6(a).

c. Calculate the value of the invariant $I(x, y)$. Is it constant in time? Compare it with figure 14.6(b).

d. Plot the solution in phase space. Is it a closed orbit? Compare it with figure 14.6(c).

As we would expect from an accurate numerical integration, the symplectic Euler integration scheme gives some fairly periodic solutions (figure 14.6(a)), a constant invariant $I(x, y)$ without drift (figure 14.6(b)), and a fairly periodic trajectory in phase space (figure 14.6(c)).

Exercise 14.6. Symplectic versus linearized solutions of the Lotka–Volterra model. Chose some parameters for the Lotka–Volterra model with initial conditions (x_0, y_0) close to the equilibrium point.

a. Using the symplectic Euler integration scheme, obtain a numerical trajectory. Numerically estimate the period of the trajectory.

b. Generate the analytical solution of the linearized system, using the code developed for exercise 14.2.

c. Compare the numerical solution with the analytical solution of the linearized model. In addition, compare their respective periods. Compare them with figure 14.7.

d. How do these results depend on the distance between the initial conditions and the equilibrium point? Can you see how the solutions given by the two methods become more similar the closer they get to the equilibrium point?

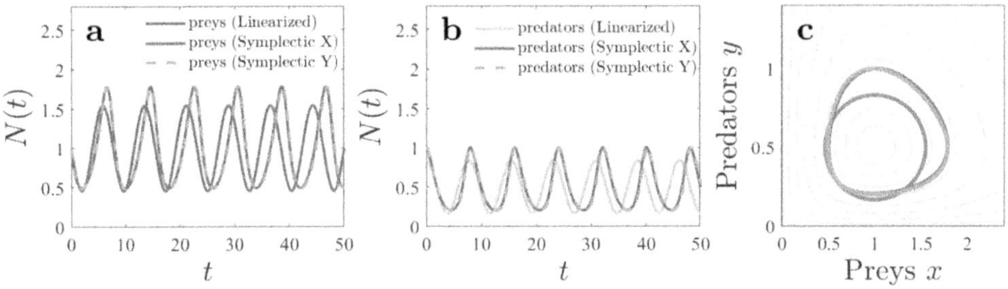

Figure 14.7. Symplectic versus linearized solutions of the Lotka–Volterra model. Comparison of the numerical solutions of the original Lotka–Volterra system (equations (14.1)), using the symplectic Euler integration schemes (equations (14.13) and (14.14)) and the linearized system (equations (14.5)) (a) for the prey, (b) for the predators, and (c) in phase space. Note that the period of the solution of the linearized system differs slightly from that of the symplectic solution of the original nonlinear system. Moreover, there is very little difference between the numerical solutions provided by the two different symplectic schemes.

14.2 The logistic growth model

The Lotka–Volterra model considers two populations (prey and predators) that interact so that they reach a dynamic equilibrium, in which neither species goes extinct or exhibits exponential uncontrolled growth. The exponential growth of a population can also be prevented by intraspecific competition, i.e., the competition between individuals of the same species for access to limited resources. The first model for this situation, proposed by Pierre François Verhulst in 1845 [5], was the *logistic growth model*:

$$\frac{\mathrm{d}x}{\mathrm{d}t} = xr - \alpha x^2, \tag{14.15}$$

where $r > 0$ is the exponential growth rate and $\alpha > 0$ is the coefficient that accounts for intraspecific competition by decreasing the growth rate by a quantity that depends linearly on the population x. In this model, the population does not grow exponentially, but it tends asymptotically to a value $K = r/\alpha$ called the *carrying capacity*. An equivalent formulation of equation (14.15) that highlights the role played by the carrying capacity is

$$\frac{\mathrm{d}x}{\mathrm{d}t} = rx\left(1 - \frac{x}{K}\right). \tag{14.16}$$

The mathematical solution of equation (14.16) for the initial condition $x(t_0) = x_0$ can be obtained by integration by the separation of variables:

$$x(t) = \frac{K}{1 + \left(\dfrac{K}{x_0} - 1\right)e^{-r(t-t_0)}}, \tag{14.17}$$

which is a sigmoid curve. To obtain biologically meaningful results, it is necessary that $0 < x_0 < K$, so that one can observe the initial exponential growth of the

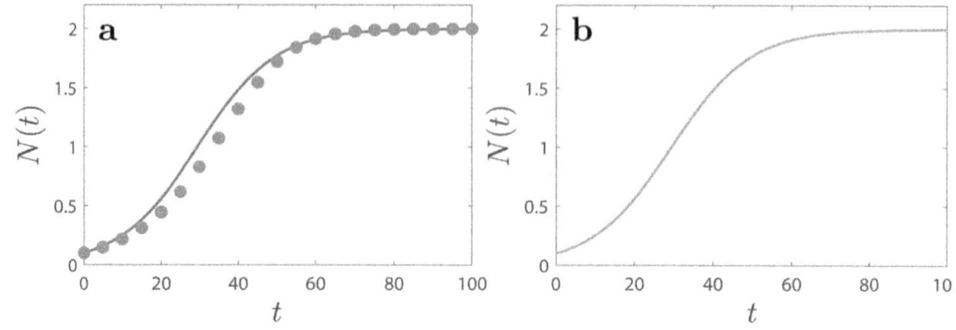

Figure 14.8. The logistic growth model. (a) Two numerical solutions of the logistic function (equation (14.15) with $r = 0.1$, $K = 2$, and $x_0 = 0.1$). The green dots are for $\Delta t = 5$, the continuous blue line is for $\Delta t = 0.05$. Both numerical solutions were obtained using an explicit Euler integration scheme. (b) Plot of the analytical solution of the logistic growth model (equation (14.17)).

population and its subsequent saturation, as shown in figure 14.8. This is a good description of population growth in confined environments, such as that of a bacteria colony in a Petri dish. (Note that, if $x_0 > K$ or $r < 0$, this equation predicts unbounded population growth—an issue first noted by Richard Levins, and consequently called *Levins' paradox*.)

Exercise 14.7. Logistic growth model.

a. Solve the logistic equation by the finite-difference method, using the explicit Euler integration method. Plot the solution obtained for different initial conditions x_0, time steps Δt, and parameters r and K. Compare the result with figure 14.8. *[Hint: Choose $\Delta t \ll r^{-1}$. Why?]*

b. Solve it using the implicit Euler integration method. Plot the solutions obtained for different x_0, r, K, and Δt. Compare them with the solution for the explicit scheme and with figure 14.8. *[Hint: First, estimate $\tilde{x}_{n+1}$ by solving the nonlinear finite-difference algebraic equation $\tilde{x}_{n+1} = x_n + r\,\tilde{x}_{n+1}\left(1 - \frac{\tilde{x}_{n+1}}{K}\right)$. Show that the estimate is then $\tilde{x}_{n+1} = K\,\frac{-(1 - r\,\Delta t) + \sqrt{(1 - r\,\Delta t)^2 + 4\,r\,\Delta t\frac{x_n}{K}}}{2\,r\,\Delta t}$.]*

c. Compare these numerical solutions with the exact solution. Which method is more appropriate? Show that the two methods are substantially equivalent in the majority of cases.

14.3 Mutualism

In Nature, mutualism is a form of interaction between two species from which both species profit. Darwin described the mutualistic interaction between orchids and their pollinators: the orchids benefit from increased genetic variety for their seeds (or their dispersal over a larger territory) and the pollinators benefit from the nutritious nectar. A first model of mutualism, called *type I mutualism*, was proposed in reference [6]. The two species with mutualistic interactions have growth that is limited by intraspecific competition, according to the logistic growth model, while an additional term, which is always positive, mimicking the interaction term in the Lotka–Volterra model, contributes to the growth rate and represents the benefit of mutualism:

$$\begin{cases} \dfrac{\mathrm{d}x}{\mathrm{d}t} &= r_1 x - \alpha_1 x^2 + \beta_{12} x y \\ \dfrac{\mathrm{d}y}{\mathrm{d}t} &= r_2 y - \alpha_2 y^2 + \beta_{21} x y \end{cases} . \tag{14.18}$$

Following the analysis performed for the Lotka–Volterra equations, the system (14.18) has four equilibrium points. Three are trivial, i.e., either one or both of the species are extinct, and only one is non-trivial:

$$\begin{cases} x_{\mathrm{eq}} &= \dfrac{\alpha_1 r_1 + \beta_{12} r_2}{\alpha_1 \alpha_2 - \beta_{12}\beta_{21}} \\ y_{\mathrm{eq}} &= \dfrac{\alpha_1 r_2 + \beta_{21} r_1}{\alpha_1 \alpha_2 - \beta_{12}\beta_{21}} \end{cases} \quad \begin{cases} x_0 &= 0 \\ y_0 &= 0 \end{cases} \quad \begin{cases} x_1 &= \dfrac{r_1}{\alpha_1} \\ y_1 &= 0 \end{cases} \quad \begin{cases} x_2 &= 0 \\ y_2 &= \dfrac{r_2}{\alpha_2} \end{cases} \tag{14.19}$$

However, in contrast to the Lotka–Volterra equations, here, the equilibrium point is unstable. This is the major drawback of this model: if it is not at equilibrium, it leads to unbounded growth.

Exercise 14.8. Type I mutualism.

a. Linearize the system (14.18) by following the same procedure as for the Lotka–Volterra system. Show that the linearized system, near the non-trivial equilibrium point, is

$$\begin{cases} \dfrac{\mathrm{d}u}{\mathrm{d}t} &= x_{\mathrm{eq}}(-\alpha_1\, u + \beta_{12}\, v) \\ \dfrac{\mathrm{d}v}{\mathrm{d}t} &= y_{\mathrm{eq}}\left(\beta_{21}\, u - \alpha_2\, v\right) \end{cases} . \tag{14.20}$$

b. Find the solution of equations (14.20). Show that this is an unstable equilibrium point. *[Hint: The first step is to find the eigenvalues λ_1 and λ_2 and eigenvectors of the matrix J:*

$$J = \begin{bmatrix} -\alpha_1\, x_{\mathrm{eq}} & \beta_{12}\, x_{\mathrm{eq}} \\ \beta_{21}\, y_{\mathrm{eq}} & -\alpha_2\, y_{\mathrm{eq}} \end{bmatrix} \equiv \begin{bmatrix} -A & B \\ D & -C \end{bmatrix} \tag{14.21}$$

where $A = \alpha_1\, x_{\mathrm{eq}}$, $B = \beta_{12}\, x_{\mathrm{eq}}$, $C = \alpha_2\, y_{\mathrm{eq}}$, and $D = \beta_{21}\, y_{\mathrm{eq}}$ are all positive quantities. Then, one builds the matrix M^{-1} of the change of base from the coordinates in u, v to the coordinates in the base of the eigenvectors e_1, e_2, and its inverse M, which gives the change of base from the base of the eigenvectors to the base of u, v. One can rewrite the system in the base of the eigenvectors e_1, e_2 using the matrix $D = M\, J\, M^{-1}$, which is a diagonal matrix with the eigenvalues λ_1 and λ_2 on the *diagonal. Therefore, one can reduce the problem of the stability to the analysis of the sign of the eigenvalues of the matrix, and on this basis, one can decide whether the solution is similar to simple harmonic motion or if it has exponential behavior.]*

c. Also linearize the system (14.18) in the proximity of the other equilibrium points. Check their stability.

d. Numerically solve the system (14.18) for different choices of parameters and initial conditions. Is the numerical solution bounded?

To avoid the inconvenience of unbounded growth as soon as the populations drift away from their equilibrium point, several strategies have been adopted to improve this model. One approach was proposed in reference [7], noting that the benefits of mutualism are not immediately available. A bound to the unlimited growth can be achieved by introducing a *handling time* T_H into the equation, i.e., the time needed to process the resources produced by the mutualistic interaction. The resulting equations are

$$\begin{cases} \dfrac{dx}{dt} = r_1 x - \alpha_1 x^2 + \dfrac{abxy}{1 + ayT_H} \\ \dfrac{dy}{dt} = r_2 y - \alpha_2 y^2 + \dfrac{abxy}{1 + axT_H} \end{cases} \tag{14.22}$$

where a is the effective search rate and b is the rate of encounters between animals of the two species. This model is more complex than type I mutualism. Analytically, these systems are difficult to analyze, due to the nature of the mutualistic term. Several studies have tried to clarify the behavior of equations (14.22) as a function of the values of their several parameters.

Exercise 14.9. Type II mutualism. Numerically solve the system (14.22) for different choices of parameters and initial conditions. Is the solution bounded?

14.4 Competition

Another situation that is often encountered in ecosystems is the competition between species that mainly rely on the same resources. A way to model this situation is inspired by the mutualistic interaction and the Lotka–Volterra equations. Competition is described using a negative interaction term that accounts for the negative effect on growth due to the simultaneous presence of the two species:

$$\begin{cases} \dfrac{dx}{dt} = r_1\, x\left[1 - \left(\dfrac{x + \alpha_{12} y}{K_1}\right)\right] \\ \dfrac{dy}{dt} = r_2\, y\left[1 - \left(\dfrac{y + \alpha_{21} x}{K_2}\right)\right] \end{cases}. \tag{14.23}$$

Following the previous line of analysis, we determine the equilibrium points. As for the mutualism, three are trivial and only one is non-trivial.

$$\begin{cases} x_{eq} &= \dfrac{K_1 - \alpha_{12}K_2}{1 - \alpha_{12}\alpha_{21}} \\ y_{eq} &= \dfrac{K_2 - \alpha_{21}K_1}{1 - \alpha_{12}\alpha_{21}} \end{cases} \quad \begin{cases} x_0 &= 0 \\ y_0 &= 0 \end{cases} \quad \begin{cases} x_1 &= K_1 \\ y_1 &= 0 \end{cases} \quad \begin{cases} x_2 &= 0 \\ y_2 &= K_2 \end{cases}. \tag{14.24}$$

As for the case of mutualism, the non-trivial equilibrium point is unstable, raising the same concerns as in that case.

Exercise 14.10. Competition. Repeat exercise 14.8 for the competition model. For each equilibrium point, derive the linearized system and its solution, and check its stability.

14.5 Further reading

For the historical manuscripts that paved the way to the study of ecosystems, the reader is referred to the original articles about the logistic model [5], the prey–predator model [3], and the paper on autocatalytic chemical reactions by Alfred J. Lotka [4] that inspired Vito Volterra. There are several review articles on this topic, for example, reference [8].

Mutualism is extremely important in ecosystems, enhancing biological diversity and survival chances for the species involved. For a review of mutualism, the reader is referred to [9]. The seminal references in which the different dynamics were proposed are reference [6] for type I mutualism and reference [7] for type II mutualism. Reference [7] discusses two possible kinds of mutualism, facultative and mandatory for a species to survive. Recent studies have been focused on finding an explanation for the evident stability of biological systems, despite their complex interaction networks. In this context, reference [9] proposes a model that is essentially a logistic model, in which the mutualistic interaction changes the growth rate r and the friction rate α.

Competition between species for the same resources is also important in ecosystems. The competition model can be generalized to N species and has been widely studied in the literature. In low dimensions, it has been proven that a limit cycle cannot occur for less than 3 species, and that chaos cannot occur for less than 4 species. Limiting cycles, stable points, and attractors can occur if 5 or more species are involved in the model. For the mathematical properties of these models, the reader is referred to [10–13]. It should be noted that a solid mathematical background is required for these references.

For more details about the explicit, implicit, and symplectic Euler methods, the reader is referred to [14].

14.6 Problems

Problem 14.1. Natural catastrophes. In a prey–predator ecosystem described by the Lotka–Volterra model, the prey is drastically reduced by an order of magnitude because of a natural catastrophic event. What are the consequences for the ecosystem?

Problem 14.2. Fluctuations. In a prey–predator ecosystem described by the Lotka–Volterra model, the prey's exponential growth rate α and the predator's exponential decline γ fluctuate in time. What are the consequences for the population dynamics if α or γ (or both) vary periodically in time? How does the period of oscillation of the parameter affect the dynamics? What if the fluctuations are random?

Problem 14.3. Ecosystem with three species. Model and simulate a system with three species: hawks, snakes, and frogs. Hawks eat snakes, snakes eat frogs, and frogs eat small insects, such as spiders, grasshoppers, and butterflies, which are available in abundance. What happens if the snakes also eat some small rodents, such as mice? And if hawks eat mice and frogs too?

Problem 14.4. Food webs. Model and simulate an ecosystem in which four or more species interact. Make these species interact using the prey–predator model in a complex food web. Study the dynamics of the resulting system.

Problem 14.5. Mutualism. Explore the dynamics of a system with four species, including two predators and two prey. Add some mutualistic interactions between some of these species.

Problem 14.6. The enrichment paradox. Read reference [15]. In this article, Michael L. Rosenzweig presents a case where, if the food supply of the prey is overabundant, the population of predators destabilizes up to the risk of extinction. Citing the abstract of the work: 'Thus man must be very careful in attempting to enrich an ecosystem in order to increase its food yield. There is a real chance that such activity may result in decimation of the food species that are wanted in greater abundance.' Simulate and discuss this system.

Problem 14.7. The Nicholson–Bailey model. The Nicholson–Bailey model is a discrete-time model that was developed to describe the population dynamics of a host–parasitoid system. If H represents the population of the host and P the population of the parasitoid, k is the reproductive rate of the host, a is the searching efficiency of the parasitoid, and c is the number of its viable eggs, the model is described by the equations

$$\begin{cases} H_{n+1} &= k\ H_n\ e^{-aP_n} \\ P_{n+1} &= c\ H_n\ (1 - e^{-aP_n}) \end{cases}.$$

Find the equilibrium points, if any, and describe the dynamics in the system. Also, simulate the system and explore its evolution over time.

Problem 14.8. Effect of measurement errors on fisheries. Assume that the growth rate of the fish population of a fishery obeys the following stochastic differential equation:

$$\frac{\mathrm{d}x}{\mathrm{d}t} = \left(R_0 + \sigma_R W(t)\right) x \left(1 - \frac{x}{K}\right) - \gamma\left(\tilde{x}, t\right),$$

where R_0 is the growth rate, W is white noise, σ_R is the standard deviation of the noise, and K is the carrying capacity. At every time step, the fishermen measure the population $\tilde{x} = x + e(t)$, where $e(t)$ is a Gaussian error, to decide their fishing quota $\gamma(\tilde{x}, t)$.

a. Show that the population should be kept around $K/2$ to optimize production. Calculate the average production and fishing rates for this optimized case.

b. Show that, if $\gamma(\tilde{x}) = \tilde{x} - K/2$, the extinction probability (i.e., probability of population reaching zero during a long simulation) might actually decrease with increasing measurement errors. *[Hint: Check reference [16].]*

14.7 Challenges

Challenge 14.1. Agent-based ecosystems. Using an agent-based approach, simulate ecosystems whose dynamics reproduce those described in this chapter. Compare the results (equilibrium points, limit cycles, stability). How does the noise intrinsic to agent-based models influence the evolution of these ecosystems when compared to the differential equation models described in this chapter?

Challenge 14.2. Noisy ecosystems. Building on the results of challenge 14.1, explore whether the addition of a noise term to the differential equations described in this chapter generates better models for the dynamics of the more realistic agent-based ecosystems.

Challenge 14.3. Alternative mutualistic models. Implement the mutualism model described in reference [9], in which the mutualistic interaction changes the growth rate r and the friction rate α of the logistic growth equation describing each species. Generalize it to more species.

References

[1] Bellard C, Cassey P and Blackburn T M 2016 Alien species as a driver of recent extinctions *Biol. Lett.* **12** 20150623

[2] Blackburn T M, Bellard C and Ricciardi A 2019 Alien versus native species as drivers of recent extinctions *Front. Ecol. Environ.* **17** 203–7

[3] Volterra V 1926 Fluctuations in the abundance of a species considered mathematically *Nature* **118** 558–60

[4] Lotka A J 1910 Contribution to the theory of periodic reactions *J. Phys. Chem.* **14** 271–4

[5] Verhulst P-F 1845 Recherches mathematiques sur la loi d'accroissement de la population *Nouv. Mem. l'Acad. R. Sci. B.-Lett. Brux.* **8** 1

[6] May R 1981 Models for two interacting populations *Theoretical Ecology. Principles and Applications* 2nd edn (Oxford: Blackwell)

[7] Wright D H 1989 A simple, stable model of mutualism incorporating handling time *Am. Nat.* **134** 664–7

[8] Mallet J 2012 The struggle for existence: How the notion of carrying capacity, *K*, obscures the links between demography, Darwinian evolution, and speciation *Evol. Ecol. Res.* **14** 627–65

[9] García-Algarra J, Galeano J, Pastor J M, Iriondo J M and Ramasco J J 2014 Rethinking the logistic approach for population dynamics of mutualistic interactions *J. Theor. Biol.* **363** 332–43

[10] Smale S 1976 On the differential equations of species in competition *J. Math. Biol.* **3** 5–7

[11] Hirsch M W 1985 Systems of differential equations that are competitive or cooperative: II. Convergence almost everywhere *SIAM J. Math. Anal.* **16** 423–39

[12] Hirsch M W 1988 Systems of differential equations which are competitive or cooperative: III. Competing species *Nonlinearity* **1** 51–71

[13] Hirsch M W 1985 Systems of differential equations which are competitive or cooperative: IV. Structural stability in three-dimensional systems *SIAM J. Math. Anal.* **21** 1225–34

[14] Heirer E, Lubich C and Wanner G 2002 *Geometric Numerical Integration—Structure Preserving Algorithms for Ordinary Differential Equations* (Berlin: Springer)

[15] Rosenzweig M L 1971 Paradox of enrichment: Destabilization of exploitation ecosystems in ecological time *Science* **171** 385–7

[16] Argun A 2016 Better stability with measurement errors *J. Stat. Phys. J. Stat. Phys.* **163** 1477–85

IOP Publishing

Aykut Argun, Agnese Callegari and Giovanni Volpe

Chapter 15

Ant-colony optimization

It has become common practice to rely on navigation devices to find the best (e.g., shortest, fastest, most environmentally friendly) route from one point to another. Instead of opening a paper map and measuring distances by means of an opisometer, we open an app on our smartphones and enter a query with origin, destination, and preferred modality of transportation. Then, in a few instants, we get a detailed travel plan.

How can such a complex task, requiring, in principle, the evaluation of a number of possible paths on the order of the factorial of the number of intersections in the area of interest, be solved in such a small amount of time by an algorithm? The incredibly high number of possibilities precludes the use of exhaustive research to list and examine all the possibilities. One of the algorithms often used to answer similar optimization problems is inspired by the dynamics of ant colonies (figure 15.1).

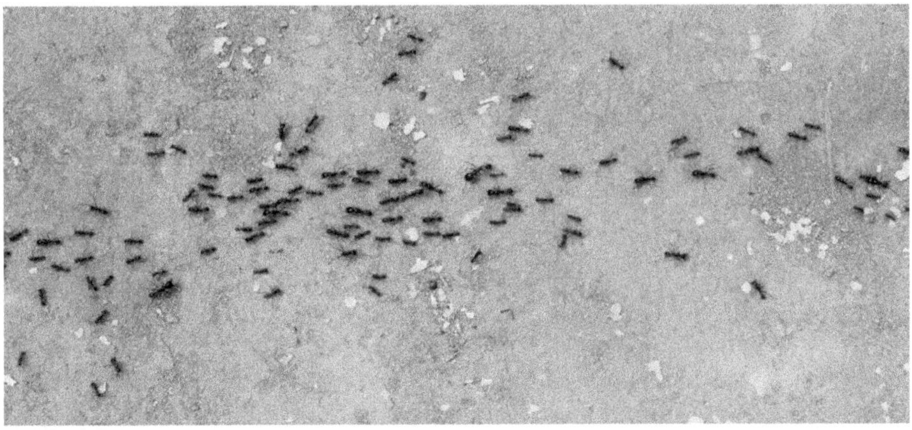

Figure 15.1. Ants following a pheromone path. Ants can find the shortest path to a food source and communicate its location to fellow ants by depositing a pheromone along the path. *Source: Pexels: Andre Moura.* http://www.pexels.com/photo/black-ants-lining-up-2563028

doi:10.1088/978-0-7503-3843-1ch15

Ants are social insects which are able to communicate very effectively. Even though individually, they are relatively simple animals, highly intelligent behavior emerges from their interaction—so-called *swarm intelligence*. For example, they explore the environment around their anthill in search of food sources. In this process, the ants appear to move randomly, but they are also partially guided by the metastable pheromone tracks left by their fellow ants. They often manage to identify the shortest path to the food source, taking into account the irregularities, obstacles, and constraints of the environment in which they are moving.

Inspired by the social behavior of ants, Marco Dorigo proposed the *ant-colony optimization algorithm* in his PhD thesis in 1992 [1], in which he addressed the problem of finding the shortest path through a graph. Since then, this method has been developed further. At present, ant-colony optimization is a class of optimization algorithms that allows us to find optimal solutions in a vast parameter space without exhaustively checking all the possibilities. Ant-colony optimization algorithms, in all their variants, have applications in the most diverse fields, from logistic and distribution networks for goods and energy, the planning of road networks, and vehicle routing and scheduling, to much less obvious applications in, e.g., protein folding, bankruptcy detection, data mining, edge detection in image analysis, and the design of antennas and nanoelectronic devices.

In this chapter, we focus on the use of original ant-colony optimization algorithm to find the shortest path in a graph. We start by introducing the problem and the tools we need for the simulation, i.e., the *graph*, the *connection matrix*, the *distance matrix*, the *weight matrix*, and the *pheromone matrix*. We guide the reader to build their own simulation that implements a rule which randomly selects the direction for the next step using the information contained in the weight matrix and the pheromone matrix, mimicking the behavior of real ants. Having implemented the algorithm, we test it in graphs of various topologies.

Example codes: Example Python scripts related to this chapter can be found at: https://github.com/softmatterlab/SOCS/tree/main/Chapter_15_Ant_Colony_Optimization. Readers are welcome to participate in the discussions related to this chapter at: https://github.com/softmatterlab/SOCS/discussions/24.

15.1 The minimum path length problem

Ant-colony optimization algorithms are applicable when the problem is stated in terms of a shortest path length (or a shortest travel time). We start by defining the environment in which our *artificial ants* are going to wander. This environment is made of a set of N spots connected by lines (figure 15.2(a)). Each spot has at least one connection, so that there are no isolated spots. Thus, this is a connected *graph* in which any point is reachable from any other point using at least one path.

The *connection matrix* M is an $N \times N$ square matrix whose elements m_{ij} are either zero or one (figure 15.2(b)). For simplicity, we consider non-directed graphs, in

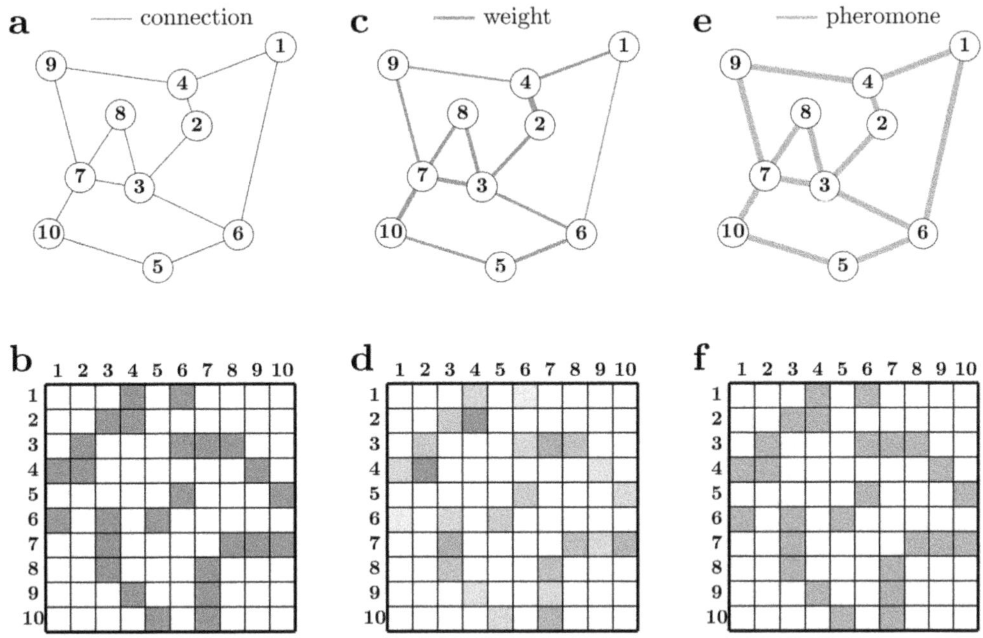

Figure 15.2. Connections, weights, and pheromone traces. Graphical representations of the (a)–(b) connections, (c)–(d) weights, and (e)–(f) pheromone traces as graphs connecting $N = 10$ points in the plane and corresponding matrices. The points and the connections can represent, e.g., cities on a map connected by roads.

which the connection matrix is symmetric: considering two vertices i and j, either $m_{ij} = m_{ji} = 0$ if there is no line connecting them, or $m_{ij} = m_{ji} = 1$ if there is a direct connection. (All the following can be generalized to the case of directed graphs with a non-symmetric connection matrix.) Also, the diagonal elements are equal to zero ($m_{ii} = 0$, i.e., the vertices are not connected to themselves by small loops).

Each connection is characterized by a distance. Let D be the *distance matrix* with elements d_{ij} (d_{ij} represents the length of the edge between nodes i and j, and should not be confused with the path length). Also, the distance matrix is symmetric (i.e., $d_{ij} = d_{ji}$). The distance used can be the geometrical distance or some other quantity. If $m_{ij} = 0$, the corresponding distance is infinite, i.e., $d_{ij} = \infty$. We also define a *weight matrix* W (figure 15.2(c)–(d)), which is usually determined using the corresponding elements of the distance matrix so that $w_{ij} = d_{ij}^{-1}$ (why? and can you think of situations where alternative definitions might be convenient?).

The problem we aim to solve is to find the path that corresponds to the minimum total distance between two vertices i_{start} and i_{end}. A path is an ordered sequence of vertices, listed one after another, e.g., $P = \{i_{\text{start}} = i_1, i_2, \dots, i_k = i_{\text{end}}\}$. The number of steps in the path is equal to the number of vertices minus one and the length of the path is

$$L(P) = \sum_{n=1}^{k-1} d_{i_n, i_{n+1}}. \tag{15.1}$$

Finally, we define the *pheromone matrix* τ (figure 15.2(e)–(f)), which is symmetric, weighted, and whose elements $\tau_{ij} = 0$ if i and j are not directly connected. The pheromone matrix will change over time; it represents the level of pheromone on a given connection between two vertices: by analogy with the signals left by real ants, our artificial ants will leave a trace of pheromone on the connections they use to reach their destinations. Only the ants that have successfully reached the destination leave the trace (on the way back to the anthill at the origin).

Exercise 15.1. Graph, connection matrix, weight matrix, paths, and distance matrix.

a. Write some code that, given N vertices, connects them randomly. Store the information about these connections in a matrix M. Plot the graph with its connections. *[Hint: Check your routine for a small number of vertices, e.g., $N = 10$. Make sure that the connection matrix is symmetric and that all diagonal elements are zero.]*

b. Generate the distance matrix using random distances. Plot it. *[Hint: Make sure it is symmetric and that the entries corresponding to non-connected vertices have infinite distance.]*

c. Generate the weight matrix from the distance matrix. *[Hint: Make sure it is symmetric and that the entries corresponding to non-connected vertices have zero weight.]*

d. Generate the pheromone matrix. *[Hint: Make sure it is symmetric and that the entries corresponding to non-connected vertices have a zero pheromone level.]*

e. Write a function that, starting from a given vertex i_1, randomly chooses a vertex i_2 directly connected to i_1, repeats the process starting from i_2, and so on for a total of $k - 1$ steps. Make sure you retrieve the information about all the vertices directly connected to a given vertex i from the connection matrix. Store the generated k-vertex path P in a sequence and return that sequence as the function's output.

f. Write a function that, given a path P, calculates the length of the path by summing the distances covered at each step, according to equation (15.1).

g. Write a function that, given a sequence of vertices, simplifies it (i.e., whenever the sequence has a repeated vertex, the subsequence between the two instances of the repeated vertex is removed together with one occurrence of the repeated vertex). *[Hint: For example: The sequence* 1, 2, 3, 2, 4 *should be simplified to* 1, 2, 4, *eliminating the jump from 2 to 3 and back. Another example:* 1, 2, 3, 5, 6, 2, 4, 5, 3 *can become either* 1, 2, 4, 5, 3 *by elimination of the subsequence between the repeated 2 or* 1, 2, 3 *by elimination of the subsequence between the repeated 3.]*

We now proceed to generate graphs with different methods, in preparation for using them in the implementation of the ant-colony optimization algorithm. We have chosen some graph topologies because they are easier to represent graphically in two dimensions without crossing edges (can you think of some graphs that do not satisfy these conditions?) so that the validity of the shortest path found can be easily verified by eye.

Exercise 15.2. Graph from Delaunay triangulation. Generate N points at random positions (x_i, y_i) with $i = 1, \ldots, N$.

- **a.** Connect the generated points by a Delaunay triangulation. Create the corresponding distance matrix M. Plot the generated graph. Compare with figure 15.3(a). *[Hint: Most software languages have built-in functions to calculate the Delaunay triangulation.]*
- **b.** Define the distance between two points i and j as their Euclidean distance (i.e., the length of the connecting segment). Generate the distance matrix D. Compare it with figure 15.3(b).
- **c.** From the distance matrix D, generate the weight matrix W.

Exercise 15.3. Graph from Voronoi tessellation. Generate N points at random positions (x_i, y_i) with $i = 1, \ldots, N$. Generate the Delaunay triangulation and the corresponding Voronoi tessellation. Consider the vertices of the Voronoi tessellation to be the vertices of the graph (note that this number will be generally different from N), and the edges of the Voronoi polygons as their connections. Repeat points **a**–**c** of exercise 15.2 for this graph. Compare the results with figures 15.3(c)–(d).

Exercise 15.4. Random square lattice. Generate a regular grid of $N_x \times N_y$ points and add a small random displacement to each point. Repeat points **a**–**c** of exercise 15.2 for this graph. Compare the results with figures 15.3(e)–(f).

Exercise 15.5. Graph on a toroidal surface. Repeat exercise 15.2 with periodic boundary conditions. Compare the results with figures 15.3(g)–(h).

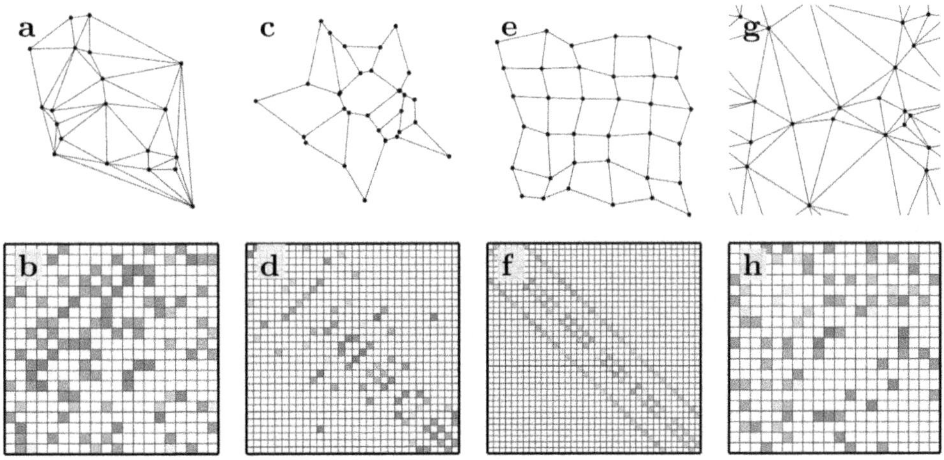

Figure 15.3. Graphs with different topologies. (a) Random points in a square connected via a Delaunay triangulation and (b) its distance matrix. The darker the shade, the closer the points. White cells indicate infinite distances, i.e., the lack of a connection between the relative points. (c) Graph obtained using a Voronoi tessellation and (d) its distance matrix. (e) An almost-square lattice, resembling the geometry of a Roman *castrum*, and (f) its distance matrix. (g) Random points in a square connected via a Delaunay triangulation with periodic boundary conditions (i.e., a graph on a toroidal surface) and (h) its distance matrix.

15.2 Ants at work

Real ants are simple social animals. They explore their environment in search of food. When a single ant finds its way to the sugar container in our kitchen cupboard, soon afterward, all its fellow ants will follow suit. How is this possible?

When an ant finds some food, it returns to the anthill by retracing its steps and leaving a pheromone track for other ants to follow. Furthermore, the ant takes into account the pheromone track already in the environment when deciding which direction to take: between two alternatives, the path with more pheromone is more likely to be taken. As more and more ants walk down a path, more and more pheromone accumulates. On the other hand, the pheromone slowly evaporates, so unused paths lose their pheromone over time and are eventually forgotten.

Coming back to our simulation with artificial ants and artificial pheromone, when an ant is at point i, it will take the direction leading to j with probability

$$p_{ij} = \frac{\tau_{ij}^{\alpha} \, w_{ij}^{\beta}}{\sum_{k} \tau_{ik}^{\alpha} \, w_{ik}^{\beta}}, \tag{15.2}$$

where $0 < \alpha < 1$ and $\beta \geqslant 1$ are two exponents that tune how the pheromone levels and weights affect the choice. (The sum at the denominator is over all possible arrival points from i; however, it can be calculated on all possible vertices without distinction, as the weights of the non-connected vertices are set to zero.)

Exercise 15.6. Branch decision rule. Code a function that implements the decision rule given by equation (15.2). Test your function on a simple graph. Make sure that the function chooses the directions with the right probabilities by testing a given branching decision over a great number of trials.

We let the ants roam around the graph deciding their next destination at each point according to rule (15.2). All ants start from vertex s_0 and aim to reach vertex f_0. When an ant reaches f_0, it stops there. We record the sequence of vertices crossed by ant n as $P_n = [s_0, i_1, i_2, \dots, i_k, f_0]$ and its length as $L(P_n) = \sum_{j \in P_n} d_{j,j+1}$. The path P_n consists of $k + 1$ steps.

After a predetermined number of steps S, for each ant n that has reached the target, we deposit an amount of pheromone equal to $Q/L(P_n)$ on each connection along its path, where Q is the total amount of pheromone that a single ant can produce to mark the path. The increment of pheromone deposited by ant n on the line connecting i and j is therefore:

$$\Delta\tau_{ij}^{[n]} = \begin{cases} \dfrac{Q}{L(P_n)} & \text{if ant } n \text{ crossed the line connecting } i \text{ and } j \\ 0 & \text{if ant } n \text{ did not cross the line connecting } i \text{ and } j \end{cases} \tag{15.3}$$

We must also take into account the evaporation of the pheromone. Given ρ, the evaporation rate, each time we update the pheromone matrix, we multiply the old matrix by the factor $(1 - \rho)$. Therefore, the rule for updating the pheromone matrix is:

$$\tau_{ij} \to (1 - \rho)\, \tau_{ij} + \sum_{n \in \{\text{ants}\}} \Delta\tau_{ij}^{[n]}, \tag{15.4}$$

where the sum only applies to ants that have reached the target.

Exercise 15.7. A first ant-colony optimization. Generate a graph with a small number N of points (e.g., $N = 40$) and choose a topology (e.g., a graph obtained from a Delaunay triangulation, as in exercise 15.2). Choose the number A of ants (e.g., $A = 20$) and decide a maximum number of steps S (e.g., $S = 80$). Choose the initial point s_0 and the destination point f_0. Make sure you set the connection matrix M, the distance matrix D, the weight matrix W, and that you initialize the pheromone matrix τ with a level of pheromone equal to one for direct connections and zero elsewhere (i.e., $\tau = M$ at the beginning). Set $\alpha = 0.8$, $\beta = 1.0$, and $\rho = 0.5$.

- **a.** Place all the ants at s_0 and let them explore the graph step by step, on the basis of rule (15.2). Keep track of each trajectory. After each step, check whether any ants have reached the target.
- **b.** After S steps, stop the simulation and plot the graph and the path of each ant that has reached its destination. Simplify these successful paths and calculate their lengths. Record the length of the shortest path l_1, and the relative sequence $P_1^{(s)}$. Compare the results to figure 15.4(a).
- **c.** Write a function that calculates the pheromone increment for a given path based on rule (15.4), and use it to update the pheromone matrix. Compare the results to figure 15.4(b).
- **d.** Run another round of S steps with the new pheromone matrix, setting the ants back to the starting point s_0. Proceed as before. After S steps, recalculate the pheromone matrix.
- **e.** Repeat many rounds of S steps. Make sure that, after the rth round, you record the length of the shortest path l_r and the relative sequence of crossed points $P_r^{(s)}$.
- **f.** After R repetitions, stop the algorithm. Plot the length of the shortest path l_r as a function of the round number, and find the minimum. Plot the paths corresponding to the minimum path length. Visually check whether the shortest path found is likely to correspond to an actual shortest path. Compare your results with the example given in figure 15.4.
- **g.** Plot the pheromone matrix at the end of the procedure, and the pheromone matrix corresponding to the first time the minimum path was found. Compare these plots (compare them also with figures 15.4(d) and 15.4(f)).

Exercise 15.8. Larger graphs. Repeat the ant-colony optimization performed in exercise 15.7 for larger graphs, e.g., take $N = 60, 120, 180, \ldots$

- **a.** Plot the length of the shortest path found in each round of S steps. How quickly is the minimum found? Is the global minimum found maintained over all the subsequent rounds?
- **b.** Do the ants explore the full graph, or is the search, in practice, restricted only to a subset of the graph, especially after a few rounds? If the ants do not run over all the possible paths, how can we be sure that we have effectively found the minimum?

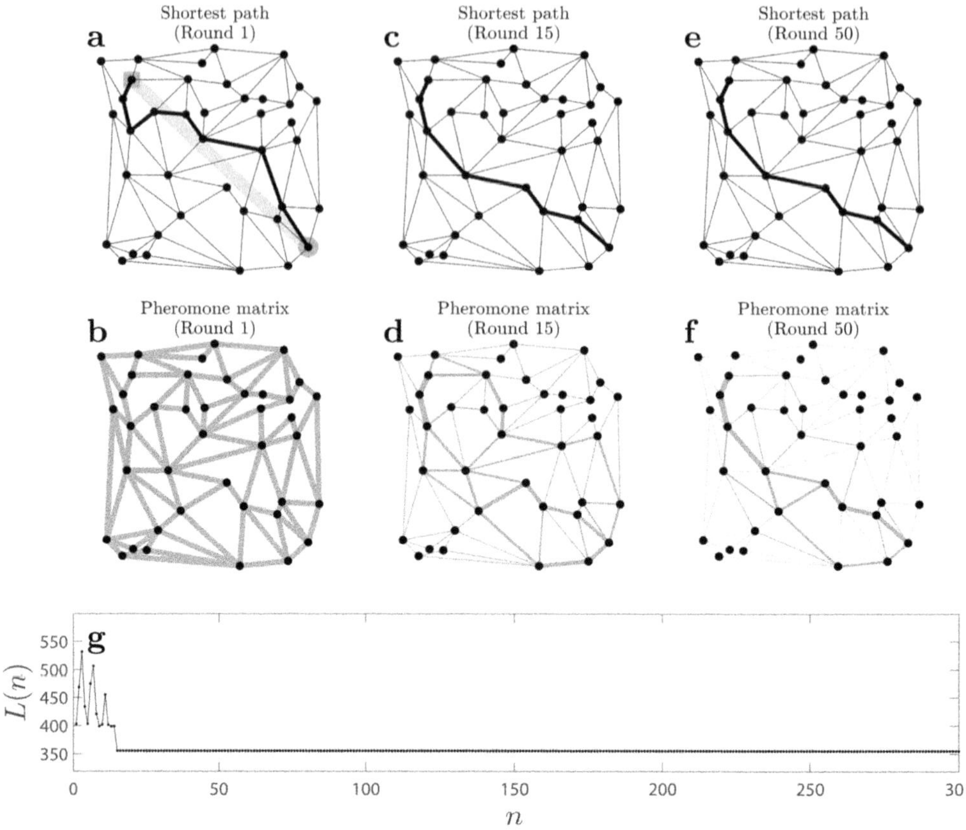

Figure 15.4. A first ant-colony optimization. (a) Graph with $N = 40$ points. The starting point is the square and the end point is the circle. The wide light blue line indicates the shortest segment connecting the points; however, it is not a connection that belongs to the graph. The shortest among the paths found during the first round (thick black line) is not the shortest possible path between the origin and the destination. (b) Pheromone matrix in the first round. (c) The absolute shortest path found in the simulation at round 15 is marked with a thick black line. (Does it look like the shortest possible path by visual inspection?). (d) Pheromone matrix when the shortest path is first found. The thicker the connections, the more pheromone they have. The distribution of the pheromone on the various segments shows that the ants are still exploring and that there is no unique established route. (e) The shortest path at round 50 still coincides with that found at round 15. (f) The corresponding pheromone matrix shows that most peripheral (and longer) connections have now been completely abandoned. (g) Length of the shortest path found at each round (300 rounds of S = 80 steps). A = 20 ants wander on the graph. The parameters for the pheromone matrix are $\alpha = 0.8$, $\beta = 1.0$, $\rho = 0.5$, $Q = 1$.

c. Try the same graph with different choices of the parameters α, β, and ρ. Do different choices give different shortest paths? Is some choice faster in finding the minimum?

d. Try the simulation using a different number of ants A walking through the graph. Does this make a difference to the algorithm? Do more ants result in a more or less accurate algorithm? What about computational efficiency?

In running the ant-colony optimization in the previous exercise, you should have noticed that the outcome of the algorithm is generally a reasonably short path, often the shortest. However, not all possible paths are explored.

From the point of view of simplicity of implementation and computation time, the ant-colony algorithm is extremely effective, especially in providing a quick answer to a query for which a good approximation of the shortest path is as desirable as the shortest path itself. This is especially valid for applications that need rerouting in real time.

Exercise 15.9. Ants on a noisy square lattice. Generate a noisy square lattice as described in exercise 15.4 (this resembles the structure of many cities founded by the Romans, with the ancient square fort structure known as the *castrum*), and proceed to implement the ant-colony optimization as described in exercise 15.8. In a perfect square lattice, there are a lot of different shortest paths, which are equivalent from the point of view of the path length. If we perturb that structure, we usually break this degeneracy, and one particular path then becomes (slightly) shorter than the others. Let us see how our ant-colony optimization algorithm performs in such a case. Compare the results with figure 15.5.

- **a.** Does the algorithm finds the shortest path? Plot and check vs the shortest path you can find by eye.
- **b.** Can you think of a possible way to improve the algorithm?
- **c.** What happens if the lattice is perfect, so that there are multiple shortest paths with exactly the same length? How is the symmetry broken?

Exercise 15.9 shows that the ant-colony optimization method finds a reasonable short path relatively fast. However, it is not always the absolute shortest path. Moreover, when the process is extended to more rounds, the shortest path is not always reinforced, and the established path at the end of the process might be a different one, which is always among the reasonable shortest paths (by direct visual inspection of the graph), but not the absolute shortest. The parameters used for the update of the pheromone matrix play a role in the determination of the established path, by affecting how fast the first-found minimum path shrinks the pheromone matrix, thereby limiting the exploration of alternative paths far away from this local minimum.

Exercise 15.10. Ants on a map from a Voronoi tessellation. Generate a Voronoi tessellation as described in exercise 15.3. Adjust the graph to obtain a map that resembles a realistic set of roads between small towns and villages in a geographical area of your liking. Choose an origin and a destination, and implement the ant-colony optimization algorithm. Compare the results with figure 15.6.

- **a.** Does the algorithm finds the shortest path? Plot and check vs the shortest path you can find by eye.
- **b.** Try different combinations of parameters for the pheromone matrix update, and determine whether there is a combination that gives faster and better results.

Exercise 15.11. Ants on a toroidal landscape. Generate a Delaunay triangulation with periodic boundary conditions as described in exercise 15.5, i.e., on a toroidal

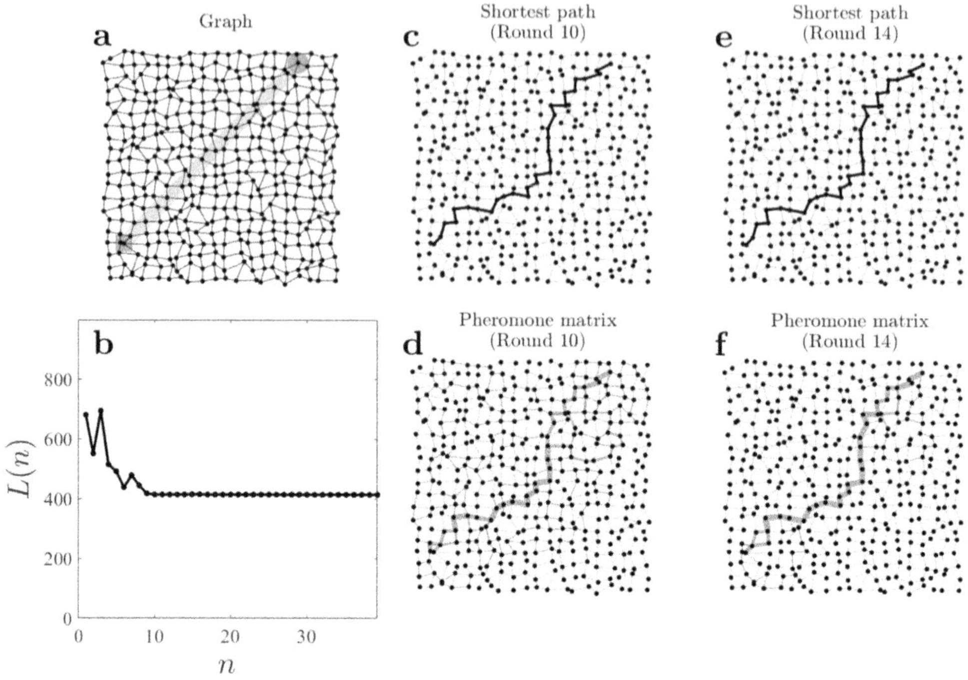

Figure 15.5. Finding the shortest path in an almost-square lattice. In this simulation, an almost-square 20 × 20 lattice is explored by 20 ants. Each round consists of $S = 800$ steps. The other parameters for the pheromone matrix update are $\alpha = 0.8$, $\beta = 1.1$, $\rho = 0.9$, and $Q = 1$. (a) A graph with the starting (square) and ending (circle) points highlighted. (b) Length of the shortest path found in each round. (c) The absolute shortest path is found at the tenth iteration. It is easy to verify by visual inspection that this is not the absolute shortest path (from the initial point, move four steps and observe that the path chosen by the ants is slightly longer then the local shortest in that area, connecting the point at step four with the point at step eight). (d) Pheromone matrix at round ten. (e) The shortest path found at round 14 is still the same, but (f) the corresponding pheromone matrix differs substantially, as many connections are now abandoned.

surface. Choose an origin and a destination, and implement the ant-colony optimization algorithm. Compare the results with figure 15.7. *[Hint: for a simple graphical representation, choose a small number of points, for instance $N = 10$.]*

15.3 Interruptions, accidents, and randomness

Let us now consider the following situation: There is a well-established route between two points, e.g., between the anthill and the sugar in the cupboard. Suddenly, one or more segments of the path become inaccessible. How long will it take the ant-colony optimization algorithm to find the new shortest path?

Exercise 15.12. Healing a disruption. Run the algorithm on a graph in the usual way, until the shortest path is found. Record the shortest path and the relative pheromone matrix when the path is found. Compare with figure 15.8.

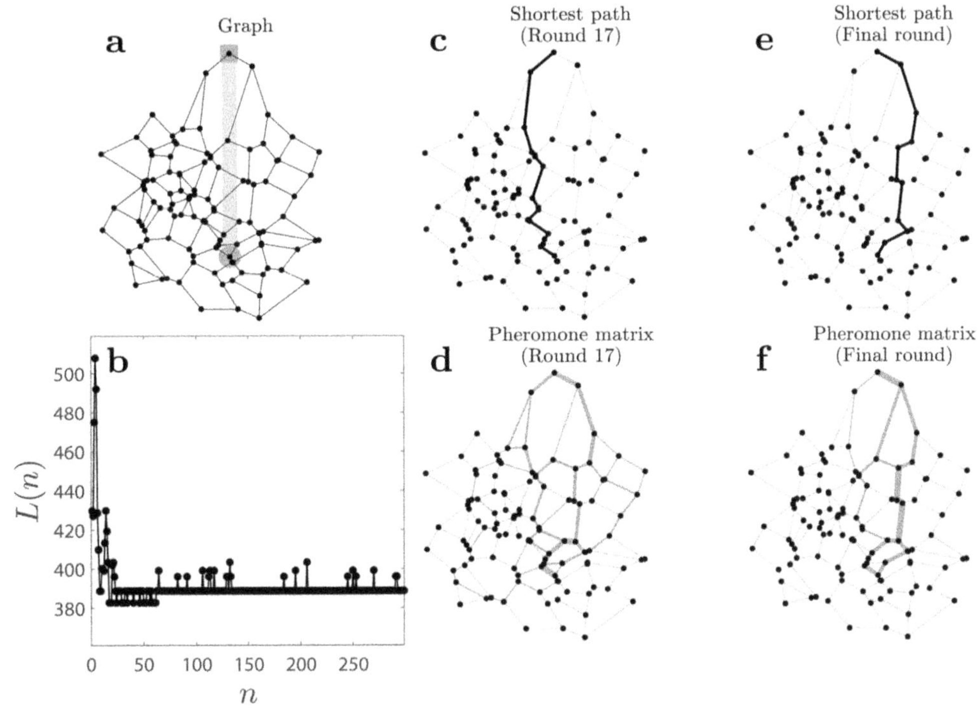

Figure 15.6. The shortest path in a Voronoi map. In this simulation, a map obtained from a Voronoi tessellation of a set of $N_0 = 60$ random points is explored by 20 ants. The resulting graph has, in this case, $N = 109$ points. Each round consists of $S = 218$ steps. The other parameters for the pheromone matrix update are $\alpha = 0.5$, $\beta = 1.1$, $\rho = 0.8$, and $Q = 1$. (a) Graph with starting (square) and ending (circle) points highlighted. (b) Length of the shortest path found at each round. (c) The shortest path of the whole simulation is first found at round 17. (d) The corresponding pheromone matrix. (e) Shortest path found in the final round (300, slightly longer than that found in round 17) and (f) the corresponding pheromone matrix.

a. Delete one or more links from the shortest path. Do this in such a way that the graph remains connected. Reflect the change in the connection matrix M, the distance matrix D, the weight matrix W, and the pheromone matrix τ.

b. Repeat the ant-colony optimization algorithm, starting from the modified matrices. What happens to the pheromone matrix after the disruption? How much time is needed to find a new shortest path?

The problem can become even more complex. While the problem of finding the shortest path in a static graph, in which the distances and connections are fixed, is mathematically a well-defined and widely investigated problem, the problem of finding the shortest path in a graph in which the connections are dynamic and can vary in time, and some stochasticity is added (e.g., in the form of terms that alter the distances or weights over time), is still lacking a thorough mathematical investigation. Therefore, even though we can certainly apply the ant-colony optimization algorithm to these kinds of problem, it is not known whether there is a solution and whether the ant-

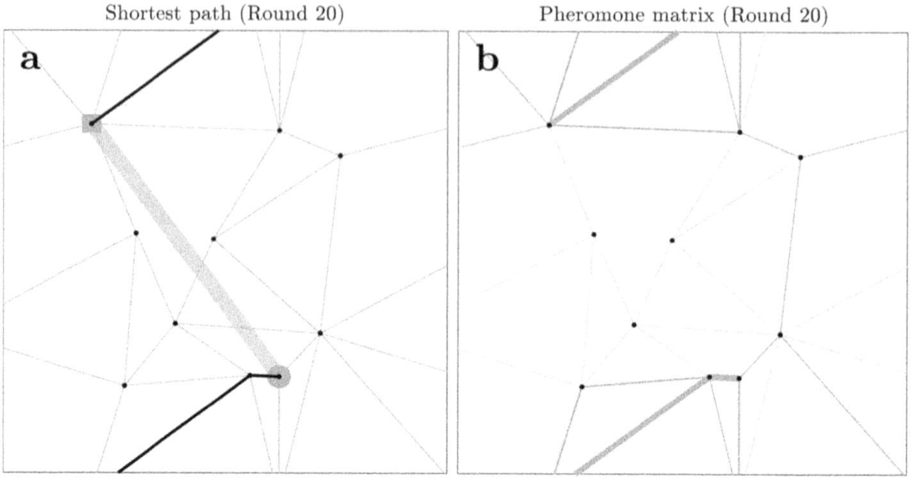

Figure 15.7. Shortest path in a toroidal world. In this simulation, a graph obtained from a Delaunay triangulation of a set of $N = 10$ random points on a toroidal geometry is explored by 20 ants. Each round consists of $S = 20$ steps. The other parameters for the pheromone matrix update are $\alpha = 0.5$, $\beta = 1.1$, $\rho = 0.8$, and $Q = 1$. (a) Graph with starting (square) and ending (circle) points highlighted. (b) Shortest path found at round 20, the final round.

colony optimization can find it in a reasonable time. To give a more concrete example, if there is a disruption to the traffic (e.g., an accident) and we ask our navigation device to find an alternative route, we might have an answer that is most likely a good one, but then if every other driver does the same, we might continue to be stuck in traffic, especially if the alternative consists of a secondary road that does not allow the same volume of traffic as the main road that is now interrupted. Then, the new path may be optimal in terms of distance, but very inconvenient in terms of time.

For even more complex situations, in which the optimization algorithm should take account of real-time data that change on a shorter timescale than the timescale required to find a solution, it is not known if the ant-colony optimization algorithm can be effective.

15.4 Further reading

The ant-colony optimization algorithm was first proposed by Marco Dorigo in his PhD thesis [1]. Several publications built on this pioneering work, e.g., references [2, 3]. It has been shown that certain variations of the ant-colony optimization algorithm are able to retrieve the global optimum in a finite time, i.e., the algorithm is convergent [2]. However, it is difficult to estimate the theoretical speed of convergence.

The problem of finding the shortest path on a graph is related to the traveling salesman problem [4] as well as to the min-delay path problem [5] and the widest path problem [6].

The ant-colony optimization algorithm and its variants have found applications in a broad range of fields. Typical problems are scheduling [7], vehicle routing [8],

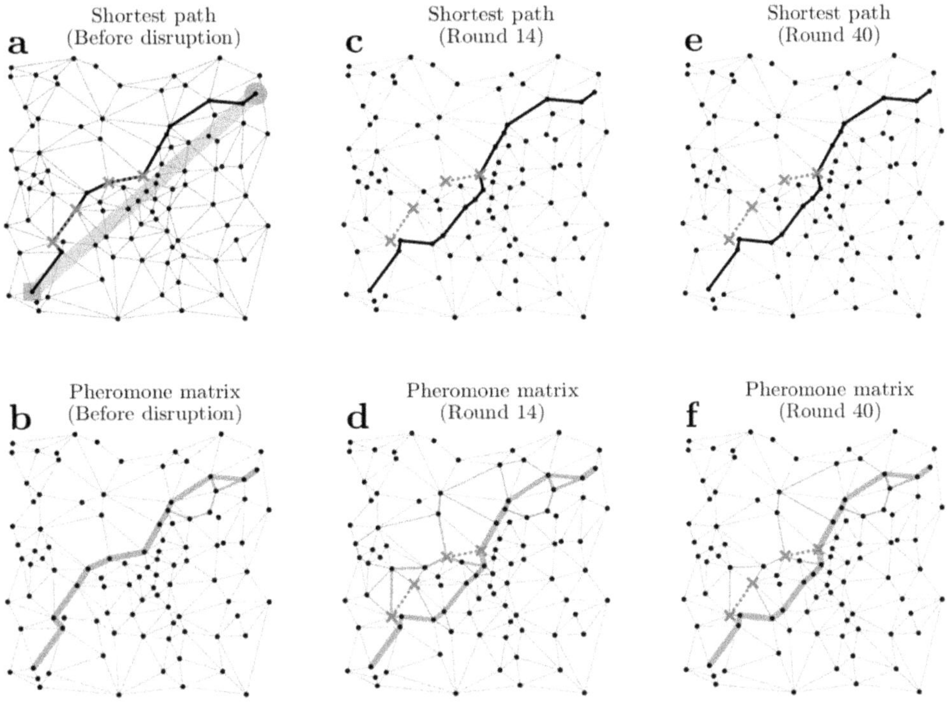

Figure 15.8. Healing a disruption. (a) Given a graph with an established shortest path (thick black line) between a start point (square) and an end point (circle), we introduce two disruptions by removing two connections (dotted red lines) and (b) relative pheromone matrix. (c) Shortest path found at round 14 and (d) corresponding pheromone matrix, which has changed from the last pheromone matrix before the disruption (notice the exploration activity picking up again). (e) Shortest path found after 40 rounds and (f) the corresponding pheromone matrix.

minimum-cost assignment [9], and set partitioning [10]. For example, they have successfully been applied to the design of nanoelectronic devices [11], to detect edges in image processing [12], to predict bankruptcy [13], to mine data [14], to classify data [15], to determine protein folding [16], to optimize energy, electricity, and telecommunications networks [17], and to plan road networks [18].

15.5 Problems

Problem 15.1. Multiple targets. Generalize the ant-colony optimization to the goal of reaching multiple targets in sequence; for example, the case in which you seek the shortest path passing first through point A, then B, then C, and so on.

Problem 15.2. Maximizing the exploration of an environment. You have a vast territory to explore. You know the number of the points on the map and the connections between the points of your map, but you do not know the exact location of the points, nor the distances between them. You send your ants (which, this time, are random walkers moving in two dimensions) to explore the environment. The ants detect a location within a certain radius from their position. Think of an algorithm to locate the majority of the points on the map. *[Hint: You can suppose that all the points are within a certain radius from the starting point.]*

Problem 15.3. Network maintenance. Servers on the internet are interconnected in a network reflecting their geographical location and are subject to periodical maintenance. They are switched off in a predefined sequence, one at each round, and switched on in the following round. Discover how information from node A reaches node B (representing two end users at two nodes that are never switched off) during each round of the maintenance schedule.

Problem 15.4. Traffic jams. A metropolitan road network is highly congested with traffic during rush hours. A vehicle starts from A to reach B, traveling at constant speed v, when the traffic flow allows it. At each intersection, the vehicle chooses its path by considering the currently-available information about traffic jams. If it chooses to go to C, then it will take the vehicle $d(A, C)/v$ time to reach C, where $d(A, C)$ is the distance between A and C. If, in the meantime, the road from A to C becomes jammed, the vehicle cannot proceed and waits until the jam is cleared. Given the probability p of having a traffic jam on each connection, find the most efficient path that can be used to go from A to B, and give an estimate of the time it will take.

Problem 15.5. Highly dynamic network. In a social network, we want information from a user to reach another user. This information must pass through a series of users who are not always online: their status stochastically switches from on to off. Apply the ant-colony optimization algorithm to this problem to find out whether the information from the initial user eventually reaches the final user, and, if so, in how much time.

15.6 Challenges

Challenge 15.1. Traffic. In a navigation a system for road vehicles, you want to implement an algorithm that takes rush-hour traffic into account and suggests alternatives to the usual shortest path. How can this be done?

Challenge 15.2. Generalization to directional graphs. Generalize the ant-colony optimization to directional graphs.

Challenge 15.3. Graphs slowly changing in time. You have to buy some groceries. It is high season in a touristic location, so there are several grocery shops, but the demand for their goods is high. You have real-time information about the amount of goods in each shop and you know the location of each shop. Because of the high demand, a shop could run out of goods at any time. You also know the locations of other customers who are on their way to buy goods. Figure out a strategy to maximize the amount of goods that you can buy.

Challenge 15.4. Optimize a path according to travel time and total distance. When you are looking at Google maps for information about an itinerary, you also find information about the travel time. Devise an algorithm to estimate and provide this kind of information to the user.

References

[1] Dorigo M 1992 Optimization, Learning and Natural Algorithms *PhD Thesis* Politecnico di Milano

[2] Zlochin M, Birattari M, Meuleau N and Dorigo M 2004 Model-based search for combinatorial optimization: a critical survey *Ann. Oper. Res.* **131** 373–95

[3] Dorigo M and Stützle T 2004 *Ant Colony Optimization* (Cambridge, MA: MIT Press)

[4] Lawler E L 1985 *The Travelling Salesman Problem: A Guided Tour of Combinatorial Optimization* (New York: Wiley)

[5] Yan S and Shih Y-L 2012 An ant colony system-based hybrid algorithm for an emergency roadway repair time-space network flow problem *Transportmetrica* **8** 361–86

[6] Pollack M 1960 *Oper. Res.* **8** 733–6

[7] Blum C 2002 ACO applied to group shop scheduling: a case study on intensification and diversification *Ant Algorithms: Proc. 3rd Int. Workshop, ANTS 2002, Brussels (Lecture Notes on Computer Science* vol 2463*)* ed M Dorigo, G Di Caro and M Sampels (Berlin: Springer) pp 14–27

[8] Toth P and Vigo D 2002 Models, relaxations and exact approaches for the capacitated vehicle routing problem *Discrete Appl. Math.* **123** 487–512

[9] Lourenço H R and Serra D 2002 Adaptive search heuristics for the generalized assignment problem *Mathware Soft Comput.* **9** 209–34

[10] Maniezzo V and Milandri M 2002 An ant-based framework for very strongly constrained problems *Ant Algorithms: Proc. 3rd Int. Workshop, ANTS 2002, Brussels (Lecture Notes in Computer Science* vol 2463*)* ed M Dorigo, G Di Caro and M Sampels (Berlin: Springer) pp 222–7

[11] Zhang J, Chung H S-H, Lo A W-L and Huang T 2009 Extended Ant Colony Optimization Algorithm for Power Electronic Circuit Design *IEEE Trans. Power Electron.* **24** 147–62

[12] Nezamabadi-pour H, Saryazdi S and Rashedi E 2002 Edge detection using ant algorithms *Soft Comp.* **10** 623–8

[13] Zhang Y, Wang S and Ji G 2013 A rule-based model for bankruptcy prediction based on an improved genetic ant colony algorithm *Math. Probl. Eng.* **2013** 753251

[14] Martens D, Baesens B and Fawcett T 2011 Editorial survey: swarm intelligence for data mining *Mach. Learn.* **82** 1–42

[15] Martens D *et al* 2007 Classification with ant colony optimization *IEEE Trans. Evol. Comput.* **11** 651–5

[16] Hu X, Zhang J, Xiao J and Li Y 2008 Protein folding in hydrophobic-polar lattice model: a flexible ant-colony optimization approach *Protein Pept. Lett.* **15** 469–77

[17] Warner L and Vogel U 2008 Optimization of energy supply networks using ant colony optimization *Proc. Environmental Informatics and Industrial Ecology (Aachen)* ed A Moeller, B Page and M Schreiber (Shaker Verlag: Aachen) pp 327–34

[18] Claes R and Holvoet T 2011 Ant colony optimization applied to route planning using link travel time predictions *2011 IEEE Int. Symp. on Parallel and Distributed Processing Workshops and Phd Forum* pp 358–65

IOP Publishing

Simulation of Complex Systems

Aykut Argun, Agnese Callegari and Giovanni Volpe

Chapter 16

The Sugarscape

A *society* is a group of individuals living together in a community, defined by geographical, economic, historical, and cultural factors. Especially in today's world, societies are not isolated and are strongly affected by interactions with other societies on a global scale. The choices of individuals greatly affect the society in which they live. Impulsive, uninformed, or irresponsible decisions might have drastic consequences for the quality of life of the society, causing economic problems, social inequalities, conflicts, and distress, and leading, in the worst case, to unstable and extreme situations. Understanding the dynamics and the implications of human decision-making is fundamental for building and consolidating a society in which individuals can benefit from living in a community and actively contribute to its progress.

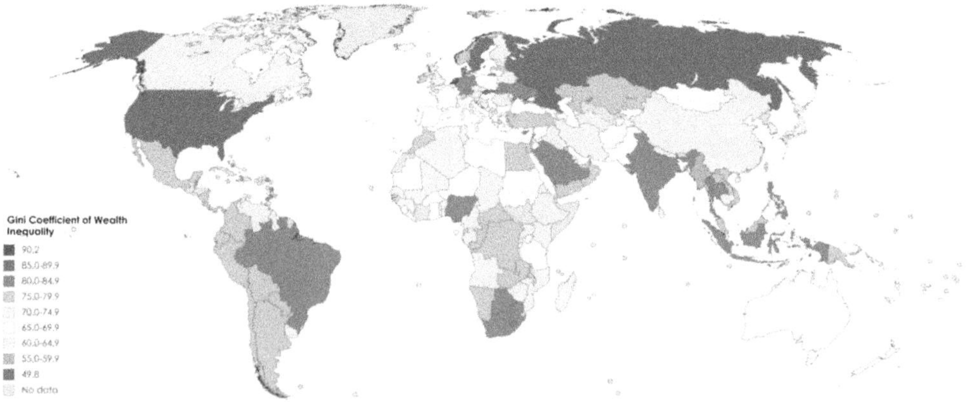

Figure 16.1. Wealth inequality in societies. Societies can have more or less wealth inequality among their population. A perfect equality scenario would be one in which every individual owns exactly the same amount as any other, while the opposite extreme would be all of the wealth in the hands of a single individual. Wealth inequality can be measured by the *Gini coefficient*, represented here in percentages for the year 2019. *Source: Wikipedia: DennisWikipediaWiki (CC BY-SA 4.0).* https://commons.wikimedia.org/wiki/File:Gini_Coefficient_of_Wealth_Inequality_source.png.

Conflicts have emerged throughout human history and also when societies were more insular and culturally homogeneous. Several additional social phenomena and challenges have emerged with the rise of more multicultural societies over the last few centuries. For example, complex racial dynamics have emerged between different ethnic groups in the United States—a phenomenon that was studied by the economist Thomas Schelling as early as the 1970s [1, 2]. His model, which presents analogies with cellular automata and Conway's Game of Life (See chapter 4), showed that segregation can spontaneously emerge in a society in which individuals have mild in-group preferences, even in the absence of any external pressure.

Twenty years after the works of Schelling, Joshua M. Epstein and Robert Axtell presented the first agent-based model for the simulation of artificial societies—the *Surgarscape* [3]. Exploiting the more advanced computational resources available, they proposed a model that, on the basis of a set of simple rules, could account for the transformation of a population through natural selection, spatial segregation, migration, and the emergence of wealth inequalities. The model could be extended by introducing additional rules to represent, e.g., trade, reproductive pressure, the exchange of information, or environment pollution. The Sugarscape, based on a set of agents acting on the basis of their individual interest, provided a new computational tool with which to study artificial societies and to predict their evolution. It has inspired a new approach in sociological predictions and analysis, and it is currently used in social sciences to complement models based on dynamical systems.

In this chapter, we first explain the model of segregation proposed by Schelling and its implications. We then explore different variations of this model and the interpretation of their prediction from a sociological point of view. We next introduce the Sugarscape and its basic rules, showing that this model features natural selection within the population, leads to the emergence of segregation, and predicts the emergence of wealth inequalities, which are observed in reality, as shown in figure 16.1. Finally, we briefly present some other rules adopted to introduce more complexity into the artificial society.

Example codes: Example Python scripts related to this chapter can be found at: https://github.com/softmatterlab/SOCS/tree/main/Chapter_16_Sugarscape. Readers are welcome to participate in the discussions related to this chapter at: https://github.com/softmatterlab/SOCS/discussions/26.

16.1 Models of segregation

We start with the model of segregation proposed by Schelling. Let us consider a town with N^2 houses organized on a square lattice with $N \times N$ cells. A fraction f of the houses are empty (and available for rent) and the remaining houses are occupied by families of two kinds, A and B, which differ by some characteristics, e.g., language, nutrition habits, or religion. Apart from the details of these differences, we know that the families discriminate on the basis that the happiness of a family depends on the proportion of families with similar characteristics living in its immediate neighborhood (a Moore's neighborhood, as defined in chapter 4). In Schelling's model of segregation, each family is happy if it is not in the minority within its neighborhood. In such conditions, Schelling showed that segregation spontaneously emerges, as we will see in the next exercise.

Exercise 16.1. Schelling's model of segregation. Consider a town with $N^2 = 2500$ houses. Suppose that the fraction of empty houses is $f = 10\%$ and that there are equal numbers of A and B families. They are initially located at random (figure 16.2(a)).

a. Implement some code that runs for 10^5 rounds. In each round, a family is chosen at random and its happiness is checked (each family is happy if it is not in the minority in its closest Moore's neighborhood, including the family itself). If it is unhappy, the family relocates to one of the free houses, chosen at random. Check your code by plotting the configuration before and after each round.

b. Show that, after a relatively small number of rounds, segregation emerges: clusters of families of the same type start to form, and the cluster size steadily increases. Compare your results with (figure 16.2(b)). Notice how the final configuration resembles the demixing of a fluid binary mixture or a phase transition of a ferromagnetic material (see chapter 2).

c. Implement a function that, given a configuration, calculates the global happiness of the whole population, and of each population fraction A and B. Show that, as time passes, the global happiness increases and the fraction of the happy population tends toward one. Compare your results with figure 16.2(c).

d. Monitor the rate of relocation events as a function of time. Show that it decreases as time passes, and tends toward zero. Compare your results with figure 16.2(c).

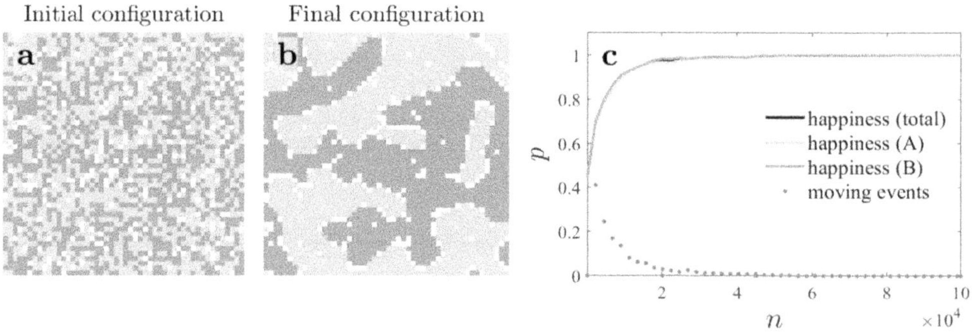

Figure 16.2. Schelling's model of segregation. Ten percent of the houses are empty (white cells); the rest are occupied by an equal number of A (yellow) and B (orange) families. Each family is happy if it is not in the minority in its closest Moore's neighborhood (including the family itself). In each round, a *random* family is chosen and, if unhappy, it relocates to a *random* empty house. (a) Starting configuration and (b) configuration after 10^5 rounds: segregation emerges, despite the randomness in the relocation process. (c) Proportion of happy families as a function of time: A (yellow), B (orange), and total (black). The total happiness increases, relocations decrease, and the pattern tends to a static limit.

The simple model proposed by Schelling lends itself to several meaningful variations. For example, one could wonder what happens if the ratio between A and B families is altered. Or, what happens if, instead of a mild gregarious spirit and a desire not to be in the minority, one or both groups strongly desire to be in the majority.

Exercise 16.2. Different levels of gregarious spirit. Perform a simulation in which the A families are happy if they are not in the minority, but the B families are happy only if they outnumber the A families by two units in a given neighborhood. Implement your model and proceed as in exercise 16.1, measuring the happiness level and visualizing the segregation level. Discuss your results.

Another interesting question is to see what happens in the case of *anti-gregarious* spirit, i.e., each family is happy when it is *not* in the majority (which can be interpreted as a desire to distinguish the family from other families, or as a desire to embrace societal diversity).

Exercise 16.3. Anti-gregarious spirit. Along the lines of exercise 16.1, simulate the dynamics for the case in which both A and B families are happy if they are not in the majority in their immediate neighborhood.

a. Compare your initial and final configurations with figures 16.3(a)–(b).

b. Calculate the total happiness and the happiness of each group as time passes and compare your results with figure 16.3(c).

c. Calculate the relocation rate and shows that it tends to zero as time passes, as the global fraction of happy families tends toward one. Compare your results with figure 16.3(c).

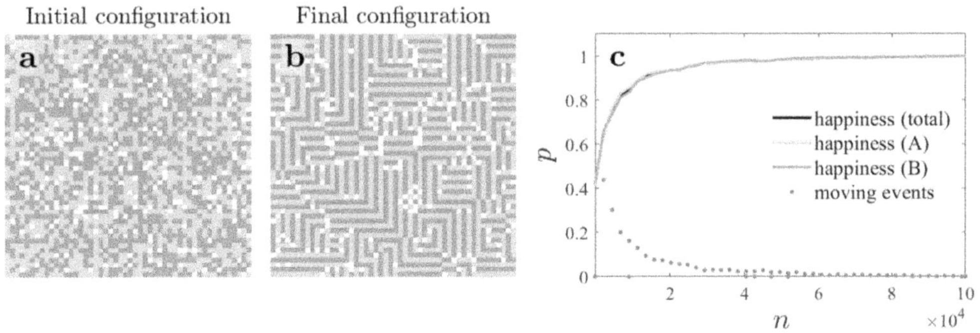

Figure 16.3. Anti-gregarious spirit. In this case, each family is happy when it is not in the majority in its neighborhood. (a) Starting configuration and (b) configuration after 10^5 moves. A clear striped pattern emerges to maximize the mixing. (c) Proportion of happy families as a function of time: type A (yellow), type B (orange), and total (black). The total happiness increases, relocations decrease, and the pattern tends to a static limit.

The striking difference between the classical model of segregation and its variation, in which each group is mildly biased toward diversity, is that, in the second case, a totally different pattern emerges: the result is a configuration in which stripes with a thickness of one alternate to favor the highest possible level of mixing.

We could push the model even further and explore what happens in the case in which the two groups have different attitudes toward in-group preference, and thus a different criterion for happiness. For example, what happens if A families are less gregarious and dislike being in the majority, but B families are more gregarious and strongly like to be in the majority in their neighborhoods?

Exercise 16.4. Emergence of frustration. Along the lines of exercise 16.1, simulate dynamics in which A families dislike being in the majority and B families want to be in the majority, exceeding by two the number of A families in their immediate neighborhood.

- **a.** Compare your initial and final configurations with figure 16.4(a)–(b).
- **b.** Calculate the total happiness and the happiness of each group as time passes and compare your results with figure 16.4(c).
- **c.** Calculate the relocation rate and show that it is approximately constant over time, reflecting the fact that the global fraction of happy families does not increase over time, and that the majority of families are unhappy. Compare your results with figure 16.4(c).

16.2 The Sugarscape

The Sugarscape model consists of a set of *agents* moving on (and interacting with) a landscape in which *sugar* grows. The sugar is a critical resource: it provides energy, wealth, and is essential to sustain the life of the agents. Each agent has an initial sugar endowment s, and at each turn, it consumes the amount of sugar dictated by its metabolic rate m. When the sugar owned by the agent is fully consumed, the

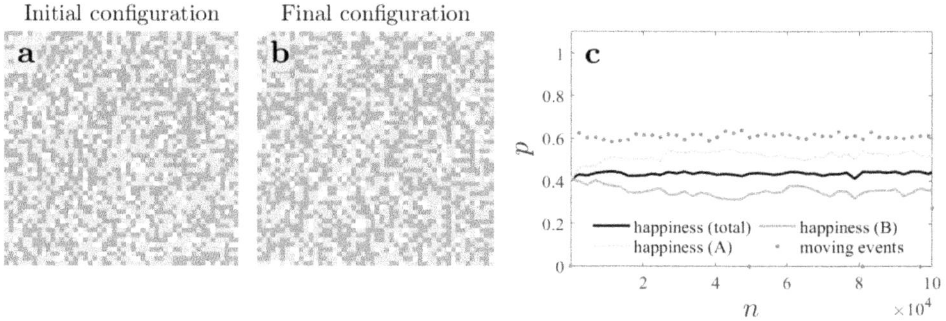

Figure 16.4. Emergence of frustration. In this case, A families are happy if they are not in the majority, while B families are happy if they are in the majority plus two. (a) Starting configuration and (b) configuration after 10^5 moves. (c) Proportion of happy families as a function of time: type A (yellow), type B (orange), and total (black). The global happiness remains constant and is far from close to one. A conspicuous fraction of the population is unhappy. Relocations are frequent and do not decrease over time.

agent dies of starvation. Therefore, the agents must explore the landscape in search of sugar: each agent has a characteristic range of vision v, which allows it to see the surrounding landscape and choose the location with the largest amount of sugar in order to harvest it.

The world in which the agents live is a square lattice of cells. Each cell contains a certain amount of sugar that might vary over time. The full amount of sugar is harvested when an agent moves into a cell, and the content of the cell drops to zero. The sugar regrows at a rate g per unit time. Each cell has a maximum amount of sugar that cannot be exceeded; therefore, if the cell is at full capacity, no more sugar can be added until its content is harvested.

In reference [3], the concepts expressed above are stated in terms of *rules*:

- rule M: concerns the movement and related actions of an agent.
- rule G_g: concerns the regrowth of sugar, where g indicates the units of sugar that regrow in each round (G_∞ indicates that the cells return immediately to their full capacity).

During each round, all the live agents move once in a random sequence. When an agent is about to move, it looks around and checks the nearest v cells that are above and below it and to the left and right. Among the cells that are not occupied by another agent, it chooses the one with the highest amount of sugar. If multiple cells satisfy this criterion, it chooses one at random. The agent then moves to that cell, harvests its sugar content, and pays the metabolic sugar amount. Therefore, if $s_n^{(i)}$ is the amount of sugar owned by agent i in its position at round n, the sugar amount owned at round $n + 1$ is

$$s_{n+1}^{(i)} = s_n^{(i)} + s_n^{(j,k)} - m^{(i)}, \tag{16.1}$$

where $s_n^{(j,k)}$ is the amount of sugar in cell (j, k) at the beginning of time step n. If $s_{n+1}^{(i)} \leqslant 0$, the agent dies and is removed from the Sugarscape. After all the agents have moved, the sugar is regrown in each cell.

Exercise 16.5. A simple Sugarscape. Take a Sugarscape of $N \times N$ cells (take $N = 50$, as in the original model) with $A = 400$ agents whose genetic features (i.e., the range of vision $v^{(i)}$ and the metabolic rate $m^{(i)}$) are chosen from a uniform distribution of integer numbers ($[v_{\min}, v_{\max}] = [1, 6]$ and $[m_{\min}, m_{\max}] = [1, 4]$). Provide each agent with an initial sugar endowment of $s_0^{(i)}$, chosen from a uniform distribution $[s_{\min}, s_{\max}] = [5, 25]$.

a. Implement some code to simulate the Sugarscape and the agents. Choose the initial sugar distribution such that it resembles the one in figure 16.5(a), i.e., with two peaks of sugar, surrounded by concentric zones with gradually less sugar. The highest peaks have cells with full capacity of four units of sugar, decreasing to three, two, and one. Smaller amounts of sugar are represented by lighter shades. White cells have no sugar content.

b. In each round, all the agents move once. The random sequence in which they move is regenerated for each round. For each agent, implement the corresponding M rule. After all agents have moved once, regenerate the sugar according to

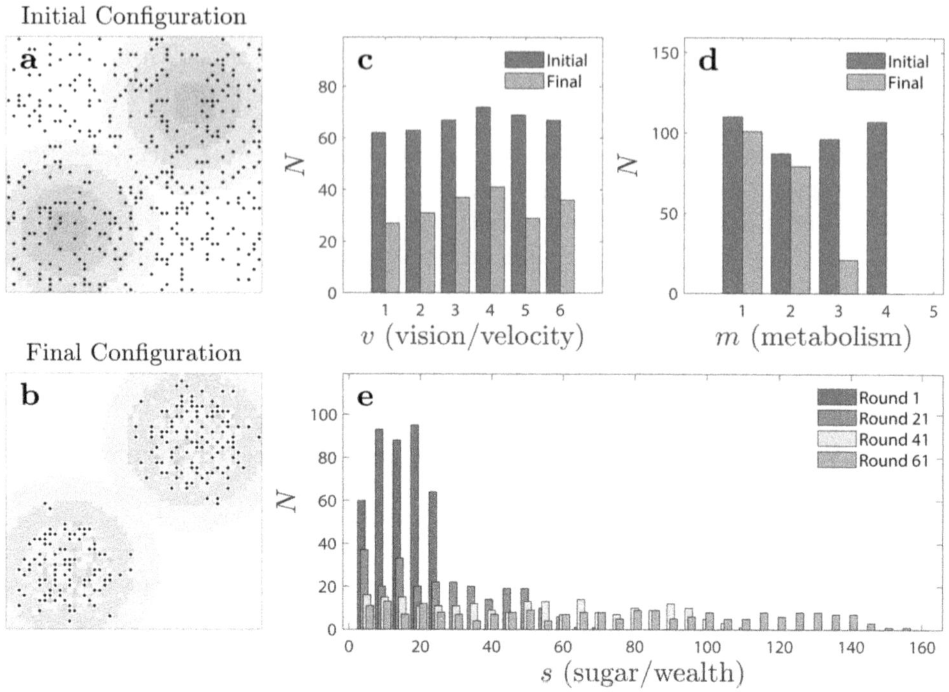

Figure 16.5. A simple Sugarscape. (a) Initial distribution of 400 agents on a Sugarscape with 50 × 50 cells (as in the original model by J. M. Epstein and R. Axtell [3]). The initial sugar content of the cells (which also corresponds to their full capacity) is represented by shades of orange: four (darkest cells), three, two, one, and zero (white cells). The sugar is replenished by one unit per round (rule G_1). (b) Final distribution of the agents after 500 rounds and the Sugarscape as it appears after the last iteration, before the last replenishment. Not all agents have survived—a fraction has died of starvation. (c) Distribution of the vision range values v among the agents (these can also be interpreted as velocity values). Initially, the distribution of the vision is almost uniform, but agents with a broader field of view tend to survive slightly better. (d) Distribution of the metabolic rate values m among the agents. Initially, the distribution of the metabolic rate values is almost uniform. However, agents with a high metabolic ratio tend to die of starvation much more than those with low metabolic consumption. As the maximum amount of sugar in a cell in this simulation is set to four, agents whose metabolic rate is equal to four go extinct in the initial rounds, and very few agents with a metabolic rate equal to three are alive at the end of the simulation. (e) Distribution of the wealth (sugar) among the population at different rounds. Initially, the maximum sugar endowment is set to 25, but as time passes, the agents remaining alive start accumulating sugar and the distribution spreads over a wider range.

the G_1 rule (i.e., in each round, one sugar unit regrows in each cell). Remove dead agents.

c. Run the simulation for a total of 500 rounds. Compare your initial and final configurations with figures 16.5(a)–(b).

d. Compare the initial and final distributions of visual ranges v and metabolic rates m. Compare your results with figures 16.5(c)–(d). Is there some genetic trait that grants a greater chance of survival?

e. Compare the distribution of wealth (sugar owned by the agents) at different times. How does it change over time? Compare your results with figure 16.5(e).

In this initial simulation, many features of the model emerge:

- **Localization/segregation**. The regions with the highest sugar capacity attract the most agents, as the spatial distribution of the agents, which is initially uniform, becomes localized in those regions.
- **Natural selection**. The genetic pool of the population changes over time, as not all genetic traits are equally favorable: agents with a high metabolic rate are more likely to die of starvation; and agents with a broader visual range are slightly favored when moving, choosing from a greater number of possible destinations.
- **Inequality in wealth distribution**. Once the population has decreased to the *carrying capacity* of the specific Sugarscape, the remaining agents are able to survive and accumulate sugar at each round. As time passes, the maximum amount of sugar owned by a single individual increases linearly with the number of iterations. Moreover, the histogram of the distribution is spread, indicating an unequal wealth distribution.

To better characterize the inequality in the wealth distribution using a more realistic simulation, we introduce an additional rule to account for a finite lifetime of the agents:

- Rule $R_{[a,b]}$: concerns the maximum age of an agent; each agent has a set maximum age $t \in [a, b]$, where a and b are integers representing number of rounds.

Initially, the age of each agent is set to zero. At each round, its age increases by one unit. An agent can die of starvation or by reaching its set maximum age. When an agent dies, a new one is generated. The newborn agent has age equal to zero, and its genetic features (v, m, t) are drawn at random from the same uniform distributions used for the initialization of the simulation. In this way, we ensure a population with a constant number of individuals, and the finite life span of the agents mimics the finite life span of humans in a society: no agent shall live forever, no unrealistic infinite accumulation of wealth happens (what if the newborn agent inherits the wealth of its dead parent?). We will use this framework to measure the degree of emergent inequality in the Sugarscape.

Exercise 16.6. Sugarscape with a finite life span for the agents. Using a Sugarscape resembling the one of exercise 16.5, implement an artificial society with 250 agents. Draw the initial maximum age of each agent from a uniform distribution in the interval [60, 100], i.e., according to the rule $R_{[60,\,100]}$. The parameters for visual range and metabolic rate are the same as in exercise 16.5.

a. Run the simulation for 500 rounds. Make sure that an agent is reborn when an agent dies. Keep track of the sugar distribution among the agents at each turn.

b. Check how the distribution of the genetic traits differs from the beginning to the end of the simulation. Compare your results with figures 16.7(a)–(b).

Before proceeding further, we must define how we are going to measure the degree of inequality in the wealth distribution. In statistics, inequality in the distribution of a resource (e.g., sugar) is usually expressed by the *Lorenz curve* or the *Gini coefficient*, which compare the actual distribution of the resource to a hypothetical equal distribution.

Let us consider a population of A individuals owning quantities $q_1 \leqslant q_2 \leqslant \cdots \leqslant q_A$ of a resource (note that we assume this to be a sequence already sorted into non-decreasing order, as shown in figure 16.6(a)). We can now calculate the cumulative sum of the resource

$$L_i = \sum_{j=1}^{i} q_j, \tag{16.2}$$

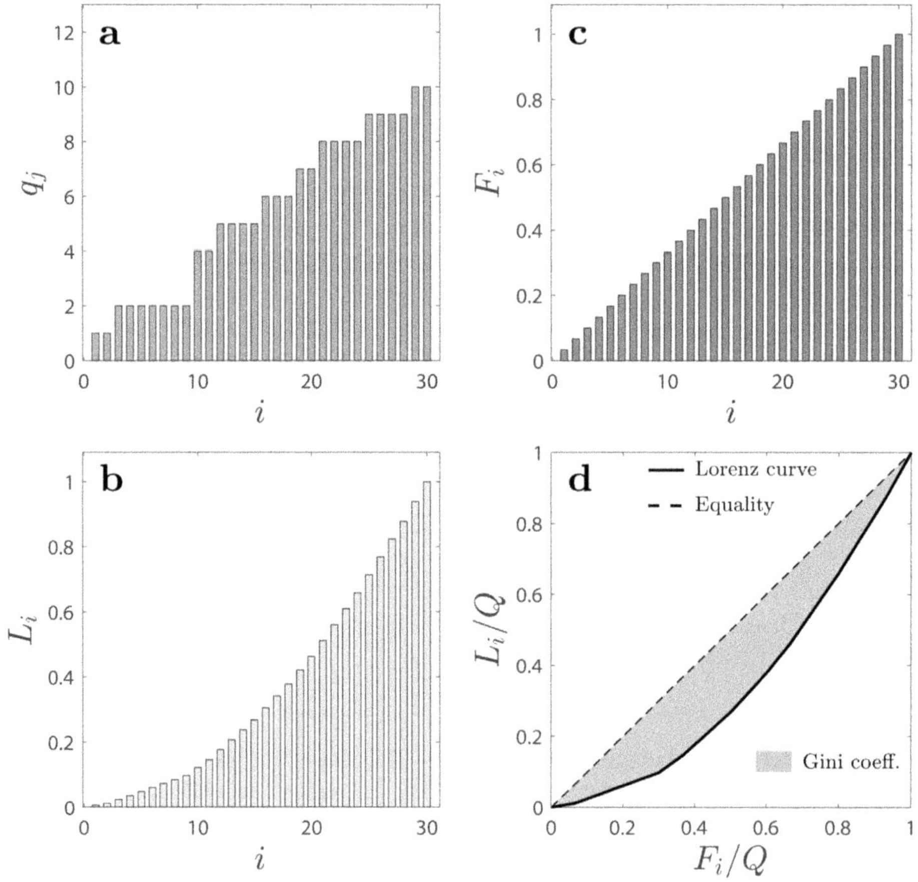

Figure 16.6. Lorenz curve and Gini coefficient. (a) Ordered sequence $\{q_i\}$ of the wealth of $A = 30$ individuals. (b) Cumulative wealth L_i owned by the individuals with ranks up to i. (c) Cumulative wealth F_i in the hypothetical scenario in which the wealth is equally divided between individuals. (d) Lorenz curve (L_i/Q versus F_i/Q where Q is the total wealth of the entire population). The Gini coefficient is twice the measure of the green area: It varies from zero (the perfect equality scenario, in which all individuals have the same wealth) to one (the highest inequality scenario, in which a single individual owns all the wealth and all the others own nothing).

as shown in figure 16.6(b). If the resource was uniformly distributed between all individuals, this function would have been

$$F_i = \frac{i}{A}Q, \tag{16.3}$$

where $Q = \sum_{j=1}^{A} q_j$ is the total amount of resource, as shown in figure 16.6(c). (What does it look like in the opposite case, in which all resources are owned by a single individual?) Then, the Lorenz curve is obtained by plotting L_i/Q as a function of F_i/Q, as shown in figure 16.6(d).

The Gini coefficient is twice the area between the Lorentz curve and the diagonal (green shade in figure 16.6(d)) and represents the amount of inequality in the distribution of the resource: the larger its value, the greater the inequality. For a perfectly uniform distribution of resources, it acquires a value of zero, and in the opposite limit, in which the resources are concentrated in the ownership of a single individual, it acquires a value of one. In economics and social sciences, it is widely used to measure the inequality in the wealth distribution of countries. In biology, it is used as a measure of the inequality of the size distribution of individuals in a population (can you guess which species have the highest and lowest Gini coefficients?).

Exercise 16.7. Lorenz curve.

a. Write a function that calculates the Lorenz curve from a set of experimental data (e_1, e_2, … , e_A) representing the individual wealth in a population. Compare your results with figure 16.6(d).

b. Show that for an equal distribution of wealth, the Lorenz curve is the straight line connecting the points (0, 0) and (1, 1).

c. Show that for the highest unequal distribution of wealth (i.e., all individuals own nothing, except for one, who owns all the resources), the Lorenz curve is a piecewise continuous line connecting the points (0, 0), (1, 0), and (1, 1).

Exercise 16.8. Gini coefficient. Write a function that numerically calculates the Gini coefficient for a set of experimental data e_j that represent the individual wealth in a population.

At this point, we are ready to analyze the inequality in the wealth distribution in the Sugarscape.

Exercise 16.9. Distribution of wealth in the Sugarscape. Building on exercise 16.6, analyze the wealth distribution by calculating the Lorenz curve.

a. Choose five different rounds (e.g., rounds 1, 21, 41, 61, and 81) and calculate the respective Lorenz curves. Compare your results with figure 16.7(c).

b. For each round n, numerically calculate the Gini coefficient $G(n)$. Plot the dependence $G(n)$ versus n. Compare the results with figure 16.7(d). Is the wealth unequally distributed in the population? Does the inequality increase over time?

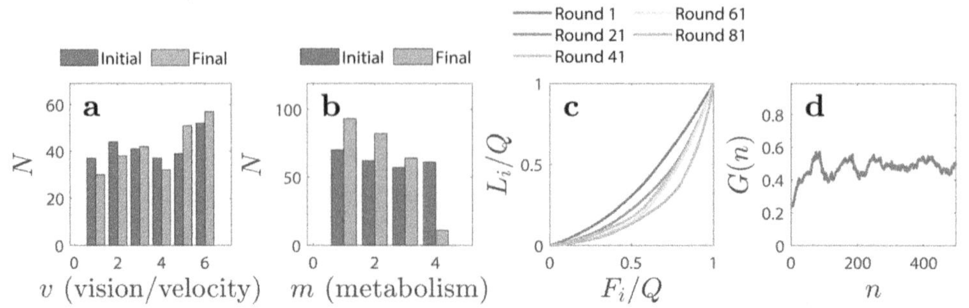

Figure 16.7. Emergence of wealth inequality. Rule $R_{[60,\ 100]}$ has now been added to the model, leading to a finite life span for the agents. (a) Initial and final distributions of the vision range, v. With time, the distribution becomes biased toward higher values of v, indicating that agents with a shorter vision range tend to die of starvation more often. (b) Initial and final distributions of the metabolic rate, m. Agents that have smaller metabolic rates are favored because they die of starvation less often. (c) Lorenz curve of the fraction of the total wealth L_i/Q versus the cumulative fraction of the wealth distributed equally F_i/Q. (d) Gini coefficient as a function of time. The Gini coefficient increases until about round 100 and then fluctuates around the value $G = 0.45$.

c. Modify the Sugarscape to introduce inheritance for the newborn agents. How does this alter the trend of the Gini coefficient?

From the simulations performed in the previous exercise, we should have obtained that, starting from an initial distribution of wealth, inequality increases as time passes, at least for the initial stage of the simulation, i.e., within the average life span of the first generation. When the agents start dying because they reach their maximum age, the Gini coefficient fluctuates over time around a value of approximately 0.45. This is because old agents die and new agents replace them. On average, the system has reached a steady state and the amount of inequality in the society is approximately constant. (How does this change if inheritance is introduced?)

The Sugarscape permits a lot of customization. Additional rules can be added to simulate different social aspects of human behavior, as was already demonstrated in reference [3]. For example, *reproductive pressure* can be added either by introducing new agents in each round, or by having agents reproduce themselves (possibly with some rules related to their age and wealth). A sort of genetic inheritance mechanism can be implemented. Sexual reproduction can also be mimicked. Complex social aspects related to the interactions between individuals have also been modeled and studied. The alternation of seasons and migratory phenomena have also been reproduced, as well as pollution associated with agent activity and sugar replenishment. Trade between individuals and loan scenarios have been investigated as well.

16.3 Further reading

The original book proposing the Sugarscape is reference [3]. The Sugarscape was inspired by the works of Thomas Schelling on the models of segregation [1, 2], which were influenced by studies of cellular automata [4] and Conway's Game of Life [5].

Robert Axelrod concomitantly developed an interesting related model showing that differences between individuals still emerge through polarization on a global scale, even in the presence of a local tendency toward convergence [6, 7]. The model presented by Axelrod is agent-based, but the agents are static, in contrast to activity in the Sugarscape. Reference [8] is an interesting article about the alignment of simulation models (i.e., the question of whether one model can include the results predicted by another model).

Several software packages have been developed to simulate the Sugarscape [9]. The best known are MASON [10], Ascape [11] and Netlogo [12]. Implementations of the Sugarscape can also be found in Mathematica [13].

Agent-based modeling is only one of the approaches used in sociology to represent human activity and human decision-making. For a broader outline of various approaches to this topic, the reader is referred to reference [14].

Some of the original results published in the book that presented the Sugarscape model [3] are difficult to replicate, as the book does not always provide the full details of the simulations. Therefore, the scientific community is currently committed to investigating how different aspects of the implementation might affect the prediction of the model. For instance, reference [15] is focused on the reproducibility question among different simulation schemes from a general point of view, and, in particular, takes the Sugarscape model and shows how different implementations of the random sequence scheme cause different limiting values in its predictions. To avoid artifacts and make robust predictions that can provide a basis for meaningful sociological interpretations, one must make sure that the results do not depend on the specific implementation of the random sequence of agent status updates.

16.4 Problems

Problem 16.1. Differently happy. In exercise 16.4 we found that when the perception of happiness is different among different groups, it might be challenging to find a configuration that satisfies the needs of everyone. Can you imagine a way to solve the frustration arising in exercise 16.4? For example, can you think of a different ratio of A and B families that would not create unhappiness?

Problem 16.2. Melting pot. What happens if, instead of two groups A and B, we have three groups A, B, and C living in the same city? Examine the different scenarios, taking into account various combinations of happiness perception, and various ratios of these groups. In addition, try this with more than three groups.

Problem 16.3. Segregation between high-income and low-income neighborhoods. In Schelling's model, add a parameter to classify the houses as expensive or cheap to rent. Also, for each family, add a parameter indicating if they are low- or high-income: A low-income (high-income) family gets some extra unhappiness when in an expensive (cheap) apartment. The relocation process happens at random, but a low-income (high-income) family should choose a cheap (expensive) house, if available. Imagine a realistic scenario, where parts of the city are cheap and others expensive.

Determine how this affects the final distribution of the families and the global happiness of the population.

Problem 16.4. Fertility and sexual reproduction in the Sugarscape. In chapter 3 of reference [3], the authors propose a way to introduce fertility and sexual reproduction in the Sugarscape. In addition to the features seen in exercise 16.9, each agent is assigned a sex. An agent is *fertile* if is it old enough (defining a childbearing age) and if it has an accumulated amount of sugar equal at least to its initial sugar endowment. For reproduction (rule S), two fertile agents of opposite sexes must find themselves in neighboring cells, and one cell in their combined neighborhood (a von Neumann neighborhood) must be free, to host the newborn child. The child has genetic features which are determined by the parents' genes, according to Mendelian rules. The sex of the child is, however, determined at random, and the initial sugar endowment is the sum of the contributions of both parents (each parent contributes half of its initial sugar endowment). Define fertility age intervals that are realistic for a human population and start with a population that has a balanced ratio of male to female individuals. The initial sugar endowment of the individuals is between 50 and 100 units. Study the evolution of the population in this scenario, taking into account the evolution of the genetic pool of the global population, also as a function of the initial population heterogeneity.

Problem 16.5. Inheritance rule. Add the inheritance rule I to the Sugarscape: when an agent dies, its accumulated wealth is equally divided among its living children. Investigate how this social convention affects the evolution of the society.

Problem 16.6. Genealogical network. The introduction of rule S (problem 16.4) naturally leads to the definition of a genealogical network. Study how the genealogical network evolves over time, also considering aspects such as spatial localization and the extent of the network. Under which conditions can you see the emergence of distinct subpopulations?

Problem 16.7. Carrying capacity of a system and the logistic equation. In a Sugarscape environment with reproduction (rule S), reproduce the dynamics leading to the logistic equation seen in chapter 14.

16.5 Challenges

Challenge 16.1. Social network in the Sugarscape. In reference [3], the authors propose a way to build a social network of agents by connecting neighbors. Visualize this social network and determine its properties (using the techniques learned in chapter 12). What happens if the movement of the agents is influenced by their social network? (E.g., if they are drawn toward/away from their higher-degree connections?)

Challenge 16.2. Culture in the Sugarscape. The genetic makeup of an agent does not change throughout its lifetime on the Sugarscape. However, there are features that do change in the life of an individual, for example, tastes. One can introduce a *cultural tag* for each agent in the Sugarscape, which is a string composed of a series of zeros and ones that simulates individual preferences or attitudes regarding some *cultural aspect* of life. The cultural tag is drawn at random for each agent at the beginning of the simulation. Cultural transmissions between agents happens as follows: when an agent has its turn, it chooses a neighbor at random. The taste of the agent regarding a specific flag (e.g., the sixth element of the cultural string) is compared to the taste of the neighbor and, in case of disagreement, transmitted to the selected neighbor. *[Hint: many transmission rules are possible; for example, the neighbor could affect the tag of the agent or the two agents could swap tags.]* Cultural groups are defined on the basis of selected positions on combinations of cultural tags. When cultural aspects are introduced into the Sugarscape, we refer to them using the short name of rule K. Study the cultural dynamics of a Sugarscape society by defining a set of cultural positions, a set of cultural groups (e.g., religion, political affiliation, musical taste) and rules for belonging to a defined cultural group. Study the evolution of the cultural aspects in time. Does one cultural group become preponderant in time, or is it rather a dialectic interaction?

Challenge 16.3. War between tribes. Having introduced cultural groups in the previous challenge, we can now define *tribes* as distinct cultural formations of agents. Tribes can come into conflict, and members of conflicting tribes can assault and prey on agents belonging to rival tribes. The combat rule is referred as rule C_α: when an assault is possible, the assaulting agent gains from the assault the minimum between α and the amount of accumulated sugar of the assaulted agent. An assault is possible when an agent sees, during its turn to move, an agent belonging to the rival tribe. The assaulting agent must be *bigger* than the assaulted agent (i.e., it must have a bigger accumulated sugar amount than its prey). More importantly, after the assault, the agent must become bigger than any other tribe fellow located in the proximity of the succumbing agent (i.e., no near friend of the assaulted can help it in fighting the attacker). When both of these conditions are fulfilled, combat is possible. To decide whether the agent will effectively conduct an assault, it decides on the move that brings the highest reward. The cells that are occupied by agents of the rival tribes are evaluated to calculate, together with their actual content of sugar, the additional reward associated with the combat. In the case in which an assault is conducted, the assaulted agent is considered to be killed and is removed from the Sugarscape. Study the evolution of the Sugarscape society when combat is allowed.

References

[1] Schelling T 1969 Models of segregation *Am. Econ. Rev.* **59** 488–93

[2] Schelling T C 1971 Dynamic models of segregation *J. Math. Sociol.* **1** 143–86

[3] Epstein J and Axtell R 1995 *Growing Artificial Societies: Social Science from the Bottom Up (Complex Adaptive Systems)* (Washinton, DC: Brookings Institution Press)

[4] Wolfram S 1984 Cellular automata as models of complexity *Nature* **311** 419–24

[5] Gardner M 1970 Mathematical Games—the fantastic combinations of John Conway's new solitaire game 'life' *Sci. Am.* **223** 120

[6] Axelrod R 1995 The convergence and stability of cultures: local convergence and global polarization *Santa Fe Institute Working Papers* 95-03-028 https://santafe.edu/research/results/working-papers/the-convergence-and-stability-of-cultures-local-co

[7] Axelrod R 1997 The dissemination of culture: A model with local convergence and global polarization *J. Conflict. Resolut.* **41** 203–26

[8] Axtell R, Axelrod R, Epstein J M and Cohen M D 1996 Aligning simulation models: a case study and results *Comput. Math. Organ. Theory* **1** 123–41

[9] Gilbert N and Bankes S 2002 Platforms and methods for agent-based modeling *PNAS* **99** 7197–8

[10] Mason (https://cs.gmu.edu/eclab/projects/mason/)

[11] Ascape (http://ascape.sourceforge.net/)

[12] Netlogo (http://ccl.northwestern.edu/netlogo/)

[13] Zeng C Sugarscape: Agent-based modeling (https://demonstrations.wolfram.com/SugarscapeAgentBasedModeling/)

[14] Millington J D A, Wainwright J and Mulligan M 2013 Representing human decision-making in environmental modelling *Environmental Modelling: Finding Simplicity in Complexity* 2nd edn (New York: Wiley) ch18 pp 291–307

[15] Kehoe J 2016 Creating reproducible agent based models using formal methods *Int. Workshop on Multi-Agent Systems and Agent-Based Simulation, MABS 2016 (Lecture Notes in Computer Science* vol 10399*)* ed L G Nardin and L Antunes (Cham: Springer) pp 42–70

www.ingramcontent.com/pod-product-compliance
Ingram Content Group UK Ltd.
Pitfield, Milton Keynes, MK11 3LW, UK
UKHW052038301026
9729IPUK00019B/305

9 780750 338448